T0305890

Artificial Intelligence

Artificial Intelligence: An Introduction to the Big Ideas and Their Development, Second Edition, guides readers through the history and development of artificial intelligence (AI), from its early mathematical beginnings through to the exciting possibilities of its potential future applications. To make this journey as accessible as possible, the authors build their narrative around accounts of some of the more popular and well-known demonstrations of artificial intelligence, including Deep Blue, AlphaGo, and even Texas Hold'em, followed by their historical background, so that AI can be seen as a natural development of the mathematics and computer science of AI. As the book proceeds, more technical descriptions are presented at a pace that should be suitable for all levels of readers, gradually building a broad and reasonably deep understanding and appreciation for the basic mathematics, physics, and computer science that are rapidly developing artificial intelligence as it is today.

Features

- Only mathematical prerequisite is an elementary knowledge of calculus.
- Accessible to anyone with an interest in AI and its mathematics and computer science.
- Suitable as a supplementary reading for a course in AI or the History of Mathematics and Computer Science in regard to artificial intelligence.

New to the Second Edition

- Fully revised and corrected throughout to bring the material up-to-date.
- Greater technical detail and exploration of basic mathematical concepts, while retaining the simplicity of explanation of the first edition.
- Entirely new chapters on large language models (LLMs), ChatGPT, and quantum computing.

Robert H. Chen is the author of three books in English on Personal Computers, Liquid Crystal Displays, and Einstein's Relativity, and four books in Chinese on LCDs & Intellectual Property, Patents, Anglo-American Contract Law, and Technology & Copyright Law, and many scholarly articles in physics and the law. He has a Ph.D. in Space Physics and a J.D. in law and is a member of the California Bar. He divides his time between California and Taiwan with his wife and daughter.

Chelsea Chen graduated in physics and computer science from UC Berkeley and is a software development engineer at a major tech company in Silicon Valley. She presently lives in Northern California and New York City.

Chapman & Hall/CRC Mathematics and Artificial Intelligence Series

Forthcoming and Recently Published Titles:

Introduction to Lattice Algebra

With Applications in AI, Pattern Recognition, Image Analysis, and Biomimetic Neural Networks

Gerhard X. Ritter, Gonzalo Urcid

Foundations of Reinforcement Learning with Applications in Finance

Ashwin Rao, Tikhon Jelvis

Artificial Intelligence

An Introduction to the Big Ideas and their Development

Robert H. Chen, Chelsea Chen

Artificial Intelligence

An Introduction to the Big Ideas and their Development

Second Edition

Robert H. Chen and Chelsea Chen

CRC Press
Taylor & Francis Group
Boca Raton London New York

CRC Press is an imprint of the
Taylor & Francis Group, an **informa** business
A CHAPMAN & HALL BOOK

Front cover image: NicoElNino/Shutterstock

Second edition published 2025
by CRC Press
2385 NW Executive Center Drive, Suite 320, Boca Raton FL 33431

and by CRC Press
4 Park Square, Milton Park, Abingdon, Oxon, OX14 4RN

CRC Press is an imprint of Taylor & Francis Group, LLC

First edition published by CRC Press 2022

Library of Congress Cataloging-in-Publication Data
Names: Chen, Robert H., 1947- author. | Chen, Chelsea C., author.
Title: Artificial Intelligence : an introduction to the big ideas and their
development / Robert H. Chen and Chelsea C. Chen.
Description: Second edition. | Boca Raton, FL : C&H, CRC Press, 2025. |
Series: Chapman & Hall/CRC mathematics and Artificial Intelligence series |
Includes bibliographical references and index.
Identifiers: LCCN 2024014843 (print) | LCCN 2024014844 (ebook) | ISBN
9781032715964 (pbk) | ISBN 9781032732978 (hbk) | ISBN 9781003463542 (ebk)
Subjects: LCSH: Artificial intelligence.
Classification: LCC Q335 .C4845 2025 (print) | LCC Q335 (ebook) | DDC
006.3--dc23/eng/20240508
LC record available at https://lccn.loc.gov/2024014843
LC ebook record available at https://lccn.loc.gov/2024014844

ISBN: 978-1-032-73297-8 (hbk)
ISBN: 978-1-032-71596-4 (pbk)
ISBN: 978-1-003-46354-2 (ebk)

DOI: 10.1201/ 9781003463542

Typeset in Minion
by Deanta Global Publishing Services, Chennai, India

Contents

Preface to the
Second Edition

ARTIFICIAL INTELLIGENCE IS PROGRESSING as fast as a Common Crawl scours the large language model (LLM) Internet, the ten trillion parameters of ChapGPT are making quick work of almost any intellectual endeavor, and the looming specter of Artificial General Intelligence (AGI) is readying the takeover of the human house; hopefully, this Second Edition will help to prepare the reader for what seems to be the inevitable appropriation of almost all human activity, but hopefully we humans can retain some relevance in the world.

The development of AI progress began with machines taking over bovine, equine, and human physical work and progressed to doing calculations for scientific research and finally producing human-level (and beyond) understanding.

This Second Edition chronologically describes the breakthrough ideas of artificial intelligence and traces its development from early computer calculation to supercomputers, quantum computing, LLMs, GPTs, and AGIs, far surpassing human brain capabilities through the artificial neural network, the mathematics of calculus, and the massive parallel processing of huge amounts of data by ever-more powerful massively parallel chips and computers.

Preface to the First Edition

IN THIS BOOK, THE adventures in the quest for artificial intelligence (AI) are exemplified by the entertaining demonstrations of "man versus machine" competitions such as IBM Deep Blue versus Garry Kasparov and Google AlphaGo versus Lee Sedol, but the real significance of AI evolves from the human *ideas* behind the machines' *algorithms* and what the machine may be capable of in the future. Starting with mechanical calculation, the development of artificial intelligence can be seen as a natural progression of technology abetted by the generation of computer science. With the hardware, software, and communications in hand, the quest for the long-dreamed of "expert system" began in earnest with the pure logic-based "top-down" machine where axioms go in and mathematical theorems come out, but because of the inherent contradictions in pure logic and lack of data and computer power, the *Logic Theorist* was replaced by Big Data and massively parallel-processing machines, in the so-called "bottom-up" approach. Bottom-up AI mimics the structure and pattern recognition capability of human brains with electrically activated artificial neurons forming synaptic patterns of recognition and "thought" in an artificial *neural network* formed in a parallel-processing computer. The synaptic patterns are produced by Markov chain modeling of neuron activation with the neurons weighted in accord with a training set in a process called *parameterization*. The gradient descent of vector calculus minimizes the difference between the machine patterns and the ground truth, backpropagating the differences using the chain rule of calculus provides the *machine learning*, and *hyperparameterization* fine-tunes the accuracy and computational efficiency of the machine's "thinking" process. The example of a convolutional neural network used in computer vision armed with Big Data and massively parallel processing has brought machine vision into almost every facet of industry and society. Predictive analytics AI is employed almost everywhere, especially critically in business, science, politics, and the military. AI reinforcement learning has produced machines that can best the top human video gamers without even *a priori* knowing the rules of the games being played, and in the real world of imperfect knowledge, AI systems have beaten the best in Texas Hold'em poker competition. Because of the ambiguities and vagaries of speech, natural language processing using *generative recurrent neural networks* that can assess immediate speech in terms of what was spoken or inferred before are presently the most promising approaches. The basic ideas behind the implementing algorithms and the processes enabling a machine to learn are described within the theme of the difference between human and artificial intelligence, with the touchstone question being the capability of machines to do mathematics. That ability required by the reader, however, is only basic calculus, and explanations of the particular equations for artificial intelligence algorithms are provided with examples so that for those who have never learned or have long forgotten their calculus, the ideas behind the algorithms can be easily understood, hopefully kindling an appreciation of the role of mathematics in artificial intelligence. As for form, some methods of mathematical exposition, for example the boldface representations of vectors and curly letter matrices, are employed only when necessary for clarity, as their identities are evident from the context. Wording and spelling are in the American style but with British punctuation leaving quotes and parentheses inside sentences of which they are only a part, at the dreadful consequence of laying bare the period or comma. Only the designs of the algorithms are presented, for those wishing to know the code, referrals are made to the relevant articles, and free code hosting platforms and excellent online programming tutorials offer great convenience and hands-on coding experience. The authors would like to thank Callum Fraser of Taylor & Francis Publishing for professional guidance, Mansi Kabra for expert handling of the manuscript, and the reviewers of the manuscript who provided excellent critical suggestions.

Prologue

H E WAS AN ORDINARY-LOOKING man with a common first and last name, Arthur Samuel, born and raised in mid-America Emporia Kansas in a middle-class family. A graduate of the local College of Emporia, he appeared to be a typical young man brought up in traditional American society. However, the genial but inherently cautious Arthur Samuel was to become known as a man of great achievement in an unusual new realm. For, from Emporia, his special talents in electronics gained him admission to the citadel of engineering education, MIT, for a master's degree in electrical engineering and upon graduation served as a lecturer there.

But World War II was raging all over the world, and after the Japanese attacked Pearl Harbor in 1941, he was recruited by the fount of technical innovation, AT&T Bell Labs, where he worked on ground-based battle-field telecommunications, and the new radar roaming the skies for enemy aircraft; both were critical to victory in the Allies' victory.

Samuel's contributions were significant, but after the War, tracking the locus of new technology, he arrived at the University of Illinois to work on the giant ILLIAC scientific computer, only to find that research funding immediately after the War was limited, and so being wooed by the cash-rich IBM, he joined in the industrial design of the world's first commercial mainframe computers.

His technical contributions were important to the computer's development, but the inherently cautious Samuel came up with an audacious marketing idea that would be the precursor of new computer applications; his idea was conceived from a return to his everyman roots as the conjurer of a "thinking machine" playing a trivial game known and played by almost everyone in America.

Working in Poughkeepsie in 1949, Arthur Samuel noted that the storage and display matrices of the IBM 701 series of new commercial computers looked just like a checkerboard, and like almost everybody in America at the time, he played that common game of jumping about a board to capture an opponents' pieces and "kinging" your own pieces.

In his uncommon mind however, Samuel saw a clever way to create an image for IBM as the creator of a wondrous computer that could challenge and defeat humans at their own game, garnering publicity for the marvelous capabilities of IBM products.

It was clearly a creative marketing scheme, but his company was not in the least supportive, for its venerable chairman Thomas Watson Sr., like Samuel, was conscious of IBM's image, but in the opposite sense of avoiding the specter of a menacing "IBM thinking machine" going around beating up on humans all over the country.

The idea of the homey IBM *Selectric* typewriters and office machines souped-up like Frankenstein's monster to triumph over small-town residents at checkers was inapposite to the friendly helpmate that was Watson Sr.'s marketing vision for IBM products.

Like Victor Frankenstein, the earnest scientist, the dogged Samuel followed his checkered muse, jumping over the IBM powers by developing the IBM 701 checkers machine on his own time. He encoded the rules of the game and programmed a *decision tree move generator* that evaluated moves by comparing numerical values of the move options that would most likely lead to a desired advantage. Included in the evaluations were nuggets of checkers wisdom familiar to any checkers player, for instance depleting the opponent's checkers by even challenges when ahead, simple gambits, and guile that today are called *heuristics*.

But this rote learning from rules, simple tree search decision evaluators, and well-known heuristics would take the 701 checkers machine only to the level of its creator, which in Samuel's case, despite his technical brilliance, was just average, so he consulted expert players for more advanced winning strategies and tactics.

Samuel found it difficult however to incorporate their skills, being mostly based on knowledge of the proclivities and idiosyncrasies of opponents, and no little hard-to-codify "feel" and outright guessing.

Since this "top-down" encoding of rules, tree searching, heuristics, and learning from some expert players, although now "expert", the 701 could only play as well as those expert players; it was clear to Samuel that any *expert system* needed to improve by playing *against* expert players, and like a human player, 701 could learn playing skills "bottom-up" and from actual game experiences, accumulate knowledge on how to win by numerically rewarding its good moves and punishing the bad moves.

Samuel now evaluated moves based upon their ultimate success or failure in training sessions based on the recorded matches of expert players, and the reward and punishment of good and bad moves in those training sessions, actual games against human players, and even in games played against itself. These were three learning schemes that would later be called, respectively, *supervised learning on training sets, reinforcement learning in game simulations,* and *self-supervised learning* by playing progressively improving versions of itself, the basic methods of today's "machine learning", a term Samuel himself coined.

By means of machine learning, the IBM 701 checkers machine slowly improved, and in 1962 after 13 years of part-time development, Samuel's machine challenged and easily defeated the fourth-ranked player in the country and Connecticut state champion Robert Nealey.

After the match, the former champion said that he "had not had such competition from anyone since 1954, when he lost his last game", but in the rueful pride characteristic of accomplished human beings, he also circumspectly exhibited an approbation of the IBM 701's "intelligence".

Any champion will say that success comes from love of the game which compels relentless practice, which requires determination. Despite its successes, the coldly logical checkers-playing IBM 701 has no love of the game, and its determination is derived from electricity, allowing endless practice every day and all night.

Samuel's IBM remained undefeated for 15 years until 1977 when it finally lost, not to a human, but to a rival checkers program developed at Duke University.

About the Authors

ROBERT **H. C**HEN IS the author of four books in English on Personal Computers, Liquid Crystal Displays, Einstein's Relativity, and Artificial Intelligence; five books in Chinese on LCDs, Intellectual Property, Patents, Anglo-American Contract Law, Technology & Copyright Law, and Artificial Intelligence; as well as many scholarly physics articles and the law in international journals such as *Science* and *Journal of Geophysical Research*. He has a Ph.D. in Space Physics and a J.D. in Law and is a member of the California Bar. He divides his time between California and Taiwan with his wife and daughter.

Chelsea Chen graduated in physics and computer science from U.C Berkeley and is a software development engineer at a major tech company in Silicon Valley. She presently lives in Northern California and New York City.

Comments on this book are welcome at robgaoxiong@gmail.com and/ or Chelseaachen96@gmail.com.

CHAPTER 1

Computing Hardware

M ANY WESTERN SCIENCE HISTORIANS have marked the beginning of the Scientific Revolution in 1543 with the publication of Copernicus' *De Revolutionibus*, an indeed revolutionary idea of the Earth revolving around the Sun, and not vice versa.

Later, Johannes Kepler's *Astronomia nova* published in 1609 presented a new ephemeris based on planetary orbits derived analytically in Newton's *Principia Mathematica* and expounded upon in Leibniz's *Specimen Dynamicum* and LaPlace's *Traité de Méchanique Celeste.*

The new scientific method of observation, hypothesis, and confirmation would enlighten and produce machines that would uplift the world's societies through modern science, and its greatest contribution was its power to predict. This power was founded on the calculus of Newton and Leibniz in the late 17th century in a new mathematics that was to be applied to scientific development and the machines that would change human society forever.

Calculus describes the rate of change of a body having position x with time t, a first derivative velocity v, and the rate of change of velocity with time that is a second derivative of position, an acceleration a, thence from Newton's second law $F = ma$, a force F applied to a body with mass m will accelerate that body thusly,

$$F = ma = m\frac{dv}{dt} = m\frac{d^2x}{dt^2}$$

One then could know precisely what force was needed to produce a desired motion, and conversely, what acceleration would produce what force.

DOI: 10.1201/9781003463542-1

Machines could be thus designed to implement those ends. Furthermore, that force F times velocity v when integrated over time t gives the amount of work W done by that force over time t and distance s, and from that, how much *Net Work* will generate how much kinetic energy (motion), and conversely how much kinetic energy is required for the machines to do that work,

$$W = \int_{t_1}^{t_2} F \cdot v\, dt = \int_{t_1}^{t_2} F \cdot \frac{ds}{dt}\, dt = \int_{s_1}^{s_2} F \cdot ds$$

and

Net Work = Kinetic Energy

The "d" in the equations denotes a very small (*infinitesimal*) change; the fundamental idea of *limits* in calculus, where, for instance, in going from point A to point B by continually halving the distance results in getting *infinitely* close to B but never arriving, the distance becoming infinitesimally small. From this concept, changes in distance, time, velocity, force, work, energy, and so on all can be made infinitesimally small, allowing representations of dynamic systems to the finest detail.

For artificial intelligence, the calculus is employed to minimize by gradient descent the error between the result of an *artificial neural network* computation, allowing the network to learn the ground truth, while the vector calculus and the chain rule of calculus pushed the network to improve its understanding through weighting parameters multiplying the artificial neurons' activation to fit the data. A system employing the calculus thus can learn and ultimately *recognize* and *infer*.

The realization of the power of calculus came in the Industrial Revolution of the late 18th century in England and was founded in large part on Robert Boyle's steam engines as the drivers of the Newtonian machines that gradually replaced human, bovine, and equine labor. Faraday's experiments and Maxwell's mathematics found that a curling magnetic field could generate an electric current in a stator, and, conversely, an electric current in the stator could produce a curling magnetic field to turn a rotor, resulting in the idea that electricity generators and driving motors could provide the energy and impetus for the machines to do the work of manufacturing and transporting.

Burning coal to boil water to produce steam to turn electromagnets to generate electricity in 1882 was scaled up by Edison and Tesla, whose great steam-powered turbines began to produce electricity not only to run machines in factories but also to light up those factories for work into the dark of night. Electricity, as well, lit up the homes in London and New York for the reading, study, and leisure that would develop an urban culture that would evolve into a new industrial society.

In 1908, Henry Ford's assembly-line mass manufacture of petroleum-powered internal combustion engines for automobile transport thrust the world into the crude oil-cracking industries of petrol and plastics that would dominate the 20th century.

The assembly line, however, relegated the erstwhile master craftsman to the mind-numbing routine of servicing the machines that now did the crafting. This not only led to labor unrest and subsequent social revolution but advanced the horrifying thought that if the machine could be intelligent as well, it would take over not only the labor but also the "thinking" from humans.

On the other hand, although man serviced the machine at work, the machine would serve man at home: the refrigerator, washing machine, and vacuum cleaner saved time and labor, and the radio, phonograph, and television provided information and entertainment.

The European and American industrial revolutions were duly noted in far-off Japan. After the 19th-century Meiji Restoration, the traditional social hierarchical order of *warrior, farmer, artisan,* and *merchant* was turned on its head by the previously unthinkable, a titled samurai, honor-bound to the rigid frugality of his spiritual class, began taking up the mundane and at times venal affairs of civil administration and the materialism of commerce.

Iwasaki Yatoro, the great-grandson of a revered samurai, in 1870 founded Japan's first *keiretsu*, Mitsubishi Heavy Industry and Shipbuilding. The builder of the agile *Zero* fighter planes and giant *Yamato* class battleships was understandably disbanded by the Americans after World War II, but upon the Korean War, to counter the rise of communism by demonstrating the virtues of capitalism, Mitsubishi was reorganized and re-branded to become an electric appliance manufacturer.

The hearts of those appliances would beat from electricity, but their soon-to-be-developed brains would depend on a new electronics based on the physics of quantum mechanical uncertainty. That is, because electrons cannot concurrently have an absolutely determinable conjugate position and energy, or conjugate time and energy, they could probabilistically tunnel through an ostensibly insurmountable potential barrier in a semiconductor material, permitting current flow and amplification under the control of an electronic gate.

From this mostly Germanic physics, first America's Bell Labs invented the transistor in 1947, and then Texas Instruments and Fairchild Semiconductor in 1958 independently created an integrated circuit of transistors and passive components, propelling TI and Fairchild's successor Intel to semiconductor device dominance in the late 20th century, and RCA, Westinghouse, and General Electric to the forefront of consumer-electronics production using those semiconductors, albeit to be quickly overtaken by the design and miniaturization wizards at Japan's Sony and Toshiba.

In nearby South Korea, after the devastation of the Korean War, under the autocratic leadership of President Park Jung-hee, the cozy relationship of central government, banks, and family-run *chaebols* fostered the dominating emergence of the Big Four of Samsung, LG, Hyundai, and Daewoo in steel, shipbuilding, consumer electronics, and finally semiconductor fabrication.

At the same time, on the small island of Taiwan, the Republic of China began textile, plastics, machine tools, and passive electronic components manufacturing for export, and the much-maligned but inexpensive and useful "Made in Taiwan" products flooded the US market, the fledgling precursors of a major electronics supply chain to mass produce semiconductors chips by TSMC, personal computers by Acer, and liquid crystal displays by Chimei, all the while bringing down prices so that people all over the world could use and enjoy the new electronics.

Japan's exquisite product design, together with South Korea and Taiwan's efficient mass production, brought high-tech products to the masses, the globalization lifting the *tiger economies* of East Asia, but drove America's pioneering consumer-electronics companies completely out of the product market.

America, however, quickly made a comeback with Texas Instruments' pocket calculator, the unprecedented computational power of IBM's mainframe computers, and Apple and IBM's personal computers, all of which together with the Asian tigers' low-cost, high-efficiency production spread the new consumer electronics all over the world.

The 20th century thus saw the trade globalization paradigm at its best: European science, American invention, Japanese design, and Korean and Taiwanese mass production. And all the while, Asia's sleeping giant, the People's Republic of China, mired in a regressive cultural revolution, was left out in the cold.

Later in the century, American research universities and the innovative spirit of Route 128 near MIT in the east and Silicon Valley near Stanford and Berkeley in the west attracted engineers and entrepreneurs from all over the world, and the new information technology companies Yahoo, Google, Facebook, and Amazon quickly rose to Internet commercial dominance, and the personal computer and smartphone began to generate the Big Data that would feed modern artificial intelligence and, on the way, shook the sleeping dragon China awake to join the fray with new-tech companies Huawei, Alibaba, Baidu, and Tencent leading the way.

The seminal changes in industry and society of the late 20th century were propelled by the computer; the middle of this century will be dominated by artificial neural networks driven by clever new algorithms parsing the ever-growing Big Data derived from the Internet and the smartphone and stored in the Cloud.

EARLY COMPUTERS

Two luminaries of the 17th-century Scientific Revolution, the French mathematician Blaise Pascal and the German scientist Gottfried Wilhelm Leibniz, designed hand-cranked cogwheel-adding machines that could also subtract and multiply by repeated addition and divide by repeated subtraction.

The teeth in the cogwheels of the *Stepped Reckoner* were the number increments, and the way of meshing the cogwheels produced the desired arithmetic process. The concept and operation of Leibniz's stepped reckoner would drive calculators for the next 300 years.[1]

In 1822, England's Charles Babbage designed a calculating machine, the *Difference Engine*, with numbers closely etched on interacting cogs and wheels such that by cranking, one could step by step iterate small differences in the independent variables dx, dt, and ds (in effect representing the infinitesimal differences in calculus), and then by adding up the function times the infinitesimals, the area underneath is just the integral as in *Work* (W) done by F in going from s_1 to s_s $W = \int_{s_1}^{s_2} F \cdot ds$, as described above. From this, the machine could compute derivatives and integrals to do calculus and even solve differential equations.

In 1834, he expanded the scope of his *Difference Engine* with a design of a locomotive-sized, steam-powered mechanical calculating device having a "store" that could hold one hundred 40-digit numbers etched on those cogs and wheels and a "mill" that could fetch the numbers and

FIGURE 1.1   The stepped reckoner, the earliest calculating machine.

perform iterations (do-loops), conditionals (if-then), and transfers (go-to), requiring many, many complex mechanical interactions. His *Analytical Engine* was the first computer that could be programmed and handle large amounts of data, the lifeblood of the artificial intelligence machines to come.

Alas, Babbage's *Analytical Engine* was never built because of the lack of funds; the world had to wait more than 100 years for Vannevar Bush's *Differential Analyzer*, still using Leibniz-like cogwheels and shafts to do the computations and set up laboriously by hand, using screwdrivers and wrenches.

Bush's mechanical *Differential Analyzer* cogwheels were gratefully replaced by Lee De Forest's vacuum-tube triode amplifiers and George Philbrick's electric voltages in 1938, thereby increasing computation speed and greatly reducing setup time. The results were displayed on an oscilloscope, signaling the arrival of the dynamic graphical representation of electronic scientific and engineering calculations.

Bush's Differential Analyzer was employed in the World War II effort, but it required ten different logical states to represent the decimals 0–9 and depended on the unstable analog voltages of hot vacuum tubes to perform its operations.

Today's digital computers' basic computation concept was recorded by George Boole in his 1847 book, *The Mathematical Analysis of Logic*, wherein he revisited Leibniz's study of ancient China's *I Ching* divination that everything under Heaven could be represented by dualities, for instance, dark and light, male and female, up and down, left and right, good and evil, and so on. Boole thus formulated a base 2 logarithmic system, thereby laying the foundations for the binary logic of the Boolean algebra that runs all computers (except quantum computing) to this day.

The basic binary logical operations of AND, OR, and NOT in combination can express all arithmetic operations, and the NOR, NAND, and XOR operations can make the logic representations simpler. Arrays and *cascades* of these *logic gates* can perform all the basic programming operations.

Logic gates receive high- or low-coded pulses through logic gate cascades to produce high and low voltages, allowing current to flow or not flow according to the truth tables of Boolean algebra to perform desired programming steps.

Implementation to modern computers came from Bush's graduate student Claude Shannon; he noted that since the binary representation of any number can be strings of binary bits, the two-state on/off logic of electronic switches could easily represent all the numbers, for example, the number 8 (binary 1000) can be easily set by four bits as binary switches as (right to left), *off, off, off, on* ($2^3 = 8$) and combinations of bits could form bytes of information indices so that combinations of switches as gates could control logical operations represented by truth tables of either True (on, open, high) or False (off, closed, low) that led to a binary logic cascade that performed the desired computation.[2]

Mathematical calculations thus could employ simple electronic switches, as Shannon outlined in his 1938 MIT master's thesis entitled "A Symbolic Analysis of Relay and Switching Circuits".

The theoretical basis for computers was set forth in Alan Turing's 1936 Cambridge University research paper, "On Computable Numbers", which defined the binary logical operations that could be sequentially recorded and stored on a theoretical infinitely long paper tape to constitute what he called a *universal machine* that could in principle compute any solvable problem and brought on the first instance of a machine-generated artificial intelligence, namely de-coding secret communications.

The deciphering of the Nazi secret code *Enigma* during World War II was done by Turing and the "Backroom Boys" at British Intelligence's Bletchley Park and their 2,000 vacuum-tube *Colossus*. Intercepted German messages were encoded and punched into paper tape which was fed to a photoelectric reader that scanned at a then-astonishing rate of 5,000 characters per second. The scan was then compared with the Enigma codes captured by the Polish resistance to find a match and through *Colossus'* iterative seeking computation, reveal the code. Even though the Germans scrambled messages by systematically rotating alphanumeric rotors and changing the plug settings and keys of the encoding machine three times a day, the "Backroom# Boys" broke the code, and as one of the Boys later said:[3]

I won't say what Turing did made us win the war,
but I daresay we might have lost it without him.

A *Turing Machine* running Boolean algebra on electronic circuits was constructed in 1942 at Iowa State College by John Atanasoff. The prototype's vacuum-tube logic and capacitor memory would later be cited as the prior art that in 1974 invalidated the ENIAC computer patent and so laid legal claim to be the world's first digital electronic computer.[4]

In 1890, a tabulator that read the holes in punch cards by trailing them under metal brushes and over a pool of electrically conductive mercury so that when the brushes penetrated a hole, a circuit between the brushes and the mercury was closed, producing a signal to add to a counter. Used for tabulating census data for the United States Census Bureau, the effort took only one-third the time of the last census to count, sort, and statistically analyze a population that had increased by 13 million to almost 63 million citizens.

The inventor, Herman Hollerith, would later start a machine tabulating business that many years later would grow under Thomas Watson Sr. to become the giant IBM that produced mainframe computers using the eponymous *Hollerith* punch cards to enter data and programming instructions for batch-mode computer processing.

Despite being a consummate profit–maximizing businessman, Watson Sr. was not immune to government calls to assist in the War effort and, as he put it, "making a virtue of necessity", he publicly proclaimed that IBM would never make more than a 1% profit from its government work. Watson Sr. then donated punch-card tabulators for the Army's ballistic table calculations, and, critically, when the German physicist Hans Bethe's equations for nuclear fission designs could not be solved by computers at Los Alamos, IBM provided the computing machines for the top-secret Manhattan Project development of the Atomic Bomb.

With the ever-increasing sophistication of computing machines, IBM agreed to a joint venture with Harvard University's Howard Aiken who had conceived a design for a fully automatic scientific computer capable of handling positive and negative numbers and carrying out mathematical equation calculations using fundamental mathematical functions, such as sines, natural logarithms, and so on.

The 50-feet-long Harvard Mark I was a decimal-coded monster with 3,304 relays, 500 miles of wire, and 750,000 electromechanical switches which clattered like "a roomful of old ladies knitting away with steel needles" while it crunched numbers up to 23 digits long, added three 8-digit numbers in a second, subtracted in 3/10 of a second, and multiplied in 3 seconds. Using the progression of reading data from continuous paper tape punch holes instead of separate punch cards, it could perform calculations in a single day that formerly took months.

Its first job was to calculate ballistic trajectories for the Navy, an extremely difficult problem, what with the swaying of ship-borne guns on an unstable sea. When shipped for operations, Watson Sr., the inveterate salesman, saw an opportunity to enhance IBM's image with a sleek, gleaming steel and glass casing for the Mark I, opposing Aiken's plan for an open frame exposing the workings for easier monitoring and adjustment.

In the introduction of the new machine to the press in 1944, Aiken barely mentioned IBM, and the announced "Harvard" surname for the Mark I further rankled Watson as diminishing IBM's role in the development of the machine, depriving the company of much sought-after publicity. The Harvard Mark I, born in acrimony, nonetheless performed admirably for the Navy, but personal animosity followed both of its midwives to their graves.

Meanwhile, in the land war, artillery pieces firing on the North African shore against Rommel's *Afrikakorps* were recoiling into the soft sand throwing off their aim. New firing tables were urgently needed. The War Department's Ballistic Research Laboratory at the Aberdeen Proving Grounds found revising the calculations beyond their differential analyzers' capabilities, so a branch was set up at the University of Pennsylvania, and the Army allocated the Moore School of Electrical Engineering $400,000 to do the new firing tables calculations.

The director of the effort, John Mauchly, had visited John Atanasoff at Iowa State to see his prototype computer, and in the fog of invention conception, he and Presper Eckert designed an 80-foot long, 30-ton Electronic Numerical Integrator and Computer (ENIAC), employing, like Atanasoff, capacitors for memory and vacuum tubes for logic that could add a thousand times faster than Harvard Mark I.[5]

However, when work was finally completed, ENIAC's unveiling in 1945 found a country no longer at war, obviating the immediate need for soft-terrain artillery firing tables, but it nonetheless could compute new, more accurate firing tables in 20 seconds, less than the time for the shell to reach the target.

A big-time assignment soon took the place of the firing tables; the development of the Hydrogen Bomb for the incipient Cold War required many, many calculations of controlled fusion reactions, and since ENIAC's vacuum tubes could switch a thousand times faster than the electromechanical switches of the Harvard Mark I, it was called upon to serve its country in a cold rather than a hot war.

The Hydrogen Bomb designs of Edward Teller and the Monte Carlo calculations of Stanislaw Ulam were clearly demonstrated in the obliteration of Elugelab island in the Enewetak Atoll; ENIAC was sanctified in the miasma of vaporized coral reefs in 1952, and although obsolete before "The Super" detonated, ENIAC was the first mainframe "general-purpose" computer (artillery firing to nuclear fusion), but surely not the last technology that was created to destroy, but ultimately served for the good of humankind.

The mainframe computer was now seen as benign but with a cool reverence-inducing capability operating with unnerving detachment, as described by a reporter witnessing IBM's Selective Sequence Electronic

FIGURE 1.2 Electronic numerical integrator and computer (ENIAC).

Calculator (SSEC) calculating a high-precision lunar ephemeris at IBM Headquarters in New York City in 1948:

> *There is the quiet clicking of printers, the steady shuffling of punched cards, the occasional rotation of a drum with memory tape, and a continual dance of little red lights as number-indicating tubes flick on and off in far less time than the twinkling of an eye. All else is hushed, and even the operators speak quietly in this streamlined sanctuary.*

The SSEC was one of a succession of acronymic computers, ILLIAC, IAS, MANIAC, ENIAC, EDVAC, EDSAC, UNIVAC, and BINAC, each contributing to the long march of automatic computing technology and each experiencing a heyday, but all quickly lapsing to desuetude after being outperformed by new designs and better technology. In this sense, Watson Sr. need not have lamented IBM's lost publicity for Mark I whose crude electromechanical switches became a symbol of backwardness in light of the ENIAC's fast vacuum-tube switches.

Although a seminal advance, ENIAC's shortcoming was that it was a base 10 decimal architecture, requiring 17,468 vacuum tubes that handled 100,000 pulses per second such that there were more than 1.7 billion chances of a tube failure every second. So, ENIAC of necessity ran at low voltages with great fans for air cooling, succeeding in reducing failures to one or two per week. But large-scale computing generally requires long sequential operations, so that the already tedious searches for the burned-out vacuum tubes that were arrayed in thousands of plug-in modules, once found, dismayingly meant re-starting the interrupted calculations all over again.[6]

Furthermore, ENIAC's "general-purpose" appellative was misleading, for its internal storage could only hold the numbers it needed for the calculations at hand. This meant that specific operations had to be connected within the circuitry by hand-plugging and unplugging hundreds of wires for each different computing task, a tedious, tiresome, and error-prone procedure that often took days to complete.

ENIAC's successor, the Electronic Discrete Variable Computer (EDVAC) operated by (discrete) binary logic rather than decimal 10, reduced the number of vacuum tubes and adopted the *von Neumann architecture* used at the Institute for Advanced Studies (IAS) computer that stored programs in an expanded electronic memory with automatic fetching replacing ENIAC's primitive hand-plugging.

In 1945, the bon vivant master of all technical disciplines, John von Neumann, in his famous *First Draft on the Report of the EDVAC* had divided computer operations into a *processor* that comprises a central arithmetic logic unit (ALU) and state registers, a *central control unit* with an instruction register and program counter, *memory* that stores instructions and data, external mass storage units, and input and output registers, altogether defining a serial architecture for the modern central arithmetic unit memory-fetch digital computer that is still used today in desktop and notebook computers.

One of these von Neumann machines was Cambridge University's Electronic Delay Storage Automatic Calculator (EDSAC), which was ironically completed two years before the EDVAC in 1949 because of turmoil at the Moore School over Mauchly and Eckert's belief that von Neumann was overly credited for *their* development of EDVAC.

Mauchly and Eckert left the University of Pennsylvania because of its policy that employees should not benefit financially from research performed at the University. With thoughts of striking it rich to assuage testy feelings, they formed their own company to produce a truly general-purpose Universal Automatic Computer (UNIVAC) using magnetic tape for high-speed programming and data input.

The design was sound but they lacked business acumen, and their financially distressed company was sold to Remington Rand in 1950. Mauchly and Eckert each made only about $300,000 from the sale and peripheral patent royalties.

Under the professional management of Remington Rand, UNIVAC was a commercial success as the first large-scale truly general-purpose computer. A total of 46 machines were sold to government and industry, one of which predicted the winner of the 1952 American presidential election, a feat that created an aura of machine intelligence that awed the public, in contrast to the esoteric ENIAC Monte Carlo simulations that were by law veiled in the secrecy surrounding the development of the Hydrogen Bomb.

UNIVAC's lead in commercial mainframe computer sales irked Watson Sr. He ordered IBM to accelerate the development of IBM's entry into the general-purpose computer market, just the IBM 701 that Arthur Samuel was hired for, and on the side, adapting it to play checkers.

For computer memory, Fred Williams at Manchester University developed a cathode ray tube, which beam painted binary dots and dashes on a phosphor screen that was read by a scanning electron beam that stored a distinctive current on a collector plate to represent data in memory that could be electronically *randomly accessed*. This *RAM* display allowed a machine to on-command fetch programs and data, and Watson Sr.'s IBM 701 immediately employed the Williams Tube to gain an edge on Remington Rand's UNIVAC.

For large amounts of data, An Wang's pulse-transfer controlling device is generally cited as the prototype of the *magnetic core memory*, whereby networks of ferrite cores stored binary bits specified by the circulation direction of magnetization produced by coaxial currents of opposite circular direction that provided the massive disk data storage for the IBM 704 and 705 computers to come.

A then commercially disinterested Harvard University allowed Wang to obtain personal patent rights, which he used to establish the eponymous Wang Labs in 1951 that pioneered the earliest desktop calculators, word processors, and scientific and business minicomputers.

In the late 1960s, after soundly defeating Remington Rand and fending off local rivals Control Data Corporation (CDC), Digital Equipment Corporation (DEC), and Amdahl, then Japan's NEC and Fujitsu, IBM's System/360 integrated circuit computers extended its mainframe hegemony to all corners of the world.

Off the silicon shores of Northern California, however, a sea change soon would downsize not only computers but also IBM itself.

NOTES

1. *Stepped Reckoner* image by Wilhelm Franz Meyer is in the public domain, Wikipedia Commons.
2. Bytes originally were defined as 8 bits that could represent $2^7 = 128$ alphanumeric and processing steps, but have increased to comprising 16, 32, 64, and so bits, each increase allowing finer and broader information indices. Analog signals, such as voice and music, can be digitized by sequentially sampling a small interval of the voltage signal, giving each sample a decimal value which is translated to binary form by an analog-to-digital-converter (ADC) and decoded back to analog voltages by a DAC for playback which, because of the assignment of numbers for each sound, is not subject to noise or distortion; that is why the music on digital CDs and MP3s sounds better, and videos on MP4s are clearer than that of analog tapes.
3. Quote by British mathematician I.J. Good who worked under Turing at Bletchley Park.
4. The term "digital" is sometimes used to describe a non-analog computing, either base 10 decimal or base 2 binary, and sometimes only for base 2 binary.

5. ENIAC image, anonymous photographer, US Army photo in the public domain.

6. Unlike the Atomic Bomb that depends on the fission of scarce ^{235}U或 ^{239}Pu, the power of a Hydrogen Bomb can be easily scaled up by plentiful and ubiquitous hydrogen (and processed deuterium and tritium isotopes). Scientists used the SSEC free of charge in fluid flow, nuclear physics, and optics calculations, and the lunar ephemeris tables it calculated were used to plot Apollo's course to the Moon in 1969. Reporter quote from *Revolution in Science, Time-Life Books* (1989).

CHAPTER 2

The Integrated Circuit

Progress was being made on all fronts and the computer age was looming on the horizon, but the dawn was held back by the bulky, high-voltage hot-running, failure-prone triode vacuum tubes. Although good enough for radios where a millionth-second short-circuit would have no noticeable effect, they would cause crash-worthy errors in the non-stop operations of electronic computing.

Once again, the great Bell Labs came to the rescue; originally researched for AT&T telecommunications and modeled after the familiar crystal radio sets thin wire "cat's whisker", when in contact with a crystal (usually silicon), will rectify and demodulate high-frequency alternating currents, such as radio waves, and transmit the signal, something that almost all boys with a curious bent would play with.

John Bardeen and Walter Brattain at Bell Labs recalled their youthful days and constructed a germanium three-terminal device that accepted a low-current signal to open the flow of a larger current between the two other terminals in the germanium, thereby amplifying the current by a *field effect*.

However, a surface layer of electrons was blocking the field, so they used two pieces of gold foil pushed by a small spring (the cat's whisker) onto the germanium surface and the contact allowed the current to flow, amplifying the current.

The 1947 invention of the transistor coolly took over the triode vacuum tube's work as an amplifier and switch at a fraction of the size. It was successfully employed in radios and calculators, shrinking their sizes and increasing their useful lives by orders of magnitude.

DOI: 10.1201/9781003463542-2

However, the spring-loaded contact was fragile and difficult to mass produce, so the irascible head of the Bell Labs transistor team, William Shockley, driven by his dismay at being one-upped by underlings Bardeen and Brattain, in the same year developed a more robust germanium flat-interface *junction transistor* that did away with the fragile contact.

Shockley's junction transistor was without doubt a breakthrough electronics achievement, and the beginning of modern computer design and processing; but although only 8×10^{-4} ounce was used per transistor, germanium was practically available only from coal fly-ash and as a by-product of zinc, silver, lead, and copper ore refining; its refining made it cost more per pound than gold. The $8 price of a germanium junction transistor compared to the $0.75 price of a vacuum tube, and pennies for resistors, inductors, and capacitors, inhibited its widespread use.

Although germanium has a higher electron/hole mobility, its chemical cousin silicon has a larger operational band gap, is more stable at higher temperatures, and is as abundant as the grains of sand on the beach. But pure crystalline silicon was difficult to produce, and the minority-carrier dopant injection necessary for semiconductor function was difficult because the surface of silicon became rough and brittle from the different temperature expansion coefficients of the materials in the junction transistor.

The breakthrough came from the impossible-to-pronounce Czochralski crystal-growth puller process that produced 99% pure crystalline silicon ingots just right for semiconductors.

The next advance was a stroke of pure luck: when Bell Labs tried doping the pure silicon in a hydrogen gas atmosphere, it accidentally ignited, and because of silicon's high chemical affinity for oxygen, a smooth SiO_2 (glass) coated the silicon, just the best natural insulator for a passivation layer, and through which holes could be etched by hydrogen fluoride to inject the dopants smoothly into the silicon to provide semiconduction.

Gordon Teal at Texas Instruments in 1953 engineered the Bell Labs process for industrial production and gradually lowered the cost of silicon junction transistors to $2.50, opening the door for their wider use, starting with TI's popular rugged little transistor radios and handy pocket calculators.[1]

The price of transistors, however, was still far above that of vacuum tubes, and for large-scale electronics requiring many vacuum tubes, the high per unit costs added up to inhibit transistor use for big electronics. It took the cost-overrun insouciance of the US military and NASA to propel the silicon transistors to electronic glory.

Avionics, weapons, and electronics for warfare, rockets, spacecraft, and instrumentation for spaceflight, with all their inherent engineering risks, and particularly the computers which controlled them, required electronic components that were robust, reliable, cool-running, long-lasting, low maintenance, small and light, just what the silicon junction transistor was all about.

With guaranteed government demand, the economies of scale soon lowered the unit costs and the transistor industry took off, literally from war into space and then to the consumer marketplace, where the benefits of the junction transistor were soon clear to all in the sleek miniaturization of beautiful new consumer electronics products and finally to mainframe general-use computers.

The transistorized products and their ever-increasing features, however, required ever-more complex circuitry with much greater numbers of transistors and passive components, and the ever-increasing power of the acronymic computers could be unleashed only by thousands of transistors replacing thousands of vacuum tubes.

The transistors were batch-processed by photoengrave etching on large wafers of silicon and then cut apart to produce individual transistors, only to be soldered to other components on circuit boards to be inserted into the devices. Although expertly wired and soldered by the nimble fingers of legions of young women, the sheer number of wires in the ever-smaller devices was fast becoming intractable, particularly for the estimated 500,000 soldering joints of thousands of wiring connections for the tens of thousands of transistors required for the ever-more powerful and versatile mainframe computers.[2]

The great potential of the digital computer was being literally strangled by the tangle of wiring connecting the transistors, a "tyranny of numbers" suppressing the freedom of the great new machines.

Coming to liberate the computers was not a revolutionary leader, but chefs of the Italian culinary arts. To avoid the tangled webs of wiring *spaghetti* connecting the transistor *meatballs*, one could simply make (a much smaller) *lasagna* instead.[3]

The active silicon semiconductor transistor meatballs and the connecting-wire spaghetti could be integrated into meat layer components and a layer of metal wiring pasta separated by layers of silicon dioxide cheese, as shown in Figure. 2.1.

FIGURE 2.1 Monolithic integration of spaghetti to lasagna.

The active transistors and passive resistors and capacitors were all fabricated from a single semiconductor "mesas" on a desert substrate, a *monolith* of all the different components into a single body.[4]

Thus, in 1958, with a nod to showmanship, Jack Kilby turned on his monolithic phase-shift oscillator that converted a DC into an AC signal and easily impressed his Texas Instruments bosses as they watched the straight-line voltage dramatically change to a sine wave on the in-line oscilloscope display.

Funding was immediate, but in the rush to file a patent application, the patent lawyers used a "flying wire" drawing with all the gangly wiring connections exposed, and his prototype did not properly claim any integrated wiring connections (Figure 2.2).[5]

In the same year, those flying wires were melted and seeped through tiny holes pre-etched in the SiO_2 insulating layer and embedded as a *metal layer* to form the printed circuit board for the conduction of planar semiconductor silicon *active layers* below.

Robert Noyce at Fairchild Semiconductor thus could demonstrate a two-state *flip-flop integrated circuit* ideal for computers that not only obviated the tedious soldering of thousands of wires, but with the micron distances between components and electrical signals traveling at about half the speed of light, the *flip-flop switch* heralded the age of the superfast integrated semiconductor transistor (Figure 2.3).[6]

And so the massive, hot glass-walled triode vacuum tubes faded to desuetude, replaced by the cool but fragile *point-contact* transistor, which was flattened out to form the planar junction transistor and finally consolidated into the completely integrated transistor chip.

The "tyranny of numbers" was overcome by *monolithic integration*, and the legions of skilled young women *solderers* were set free by the new technology of automated semiconductor planar-process fabrication in an oft-repeated tale of machines replacing humans.

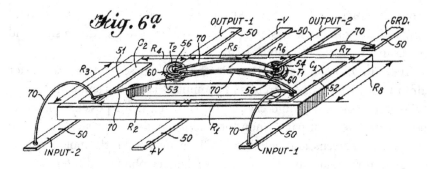

FIGURE 2.2   Kilby "flying wire" integrated circuit.

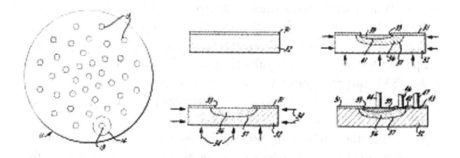

FIGURE 2.3   Noyce flip-flop integrated circuit.

The transistor switches and amplifiers were soon integrated with the passive electronic components and the integrated circuits (IC) progressively shrank in both size and cost, to become the brain cells of the modern computer. Fairchild and later Intel's *central processing units* (CPUs) would become the computing workhorses of the mainframe and personal computers.

Computer memory also used the transistor to control the reading into, and writing out of, capacitors (also in the ICs); the capacity of these *memory cells* quickly increased by orders of magnitude to become TI's *dynamic and static random access memory* (DRAM and SRAM) that came to be the information silos of the new computers.

Millions of CPUs and memory cells, at last free from their wiring bondage, were arrayed on motherboards to be inserted as complete electronic entities into electronic devices.

Invented in 1963, the soon-to-be dominant low-noise, cool-running, zero-volt static power (needs voltage only when alternating between 0 and 1), thin-gate *complementary metal-oxide semiconductor* (CMOS) that "complemented" *n*-type (NMOS) and *p*-type (PMOS) transistors could be formed under dense design rules on a wafer, bringing down sizes and prices in abeyance with Moore's Law.[7]

The CMOS would soon become the workhorse semiconductor for almost all the integrated circuits in computers large and small and in the thin-film version for liquid crystal displays that made possible the note-book computer, flat-panel monitors, wall-hanging television sets, and smartphone screens, and indeed the LCD made possible the mobile phone, which in turn made possible not only mobile telephone and Internet con-tact but also instant calculations, marketing, music, and ride-hailing, all the while (inadvertently?) amassing the Big Data and forming the Cloud for the training sets of the artificial intelligence machines to come.

In the late 1960s, after soundly defeating Remington Rand and fend-ing off local rivals Control Data Corporation (CDC), Digital Equipment Corporation (DEC), and Amdahl, then Japan's NEC and Fujitsu, IBM's System/360 integrated circuit computers extended its mainframe hege-mony to all corners of the world, and there are still an estimated 10,000 mainframes in use today, mostly from IBM.

But while the mainframe was still used in industry, government, research institutes, and universities, its dominance was co-opted by the democratization of mini and personal computers.

The tiny microprocessor could perform the basic function of an early mainframe computer; the world's first was the 2,250 transistors 4-bit Intel 4004 made for the Japanese calculator maker Busicom in 1971. A later version in 1972, the 8008 microprocessor, arranged 8-bit bytes into 256 unique arrays of ones and zeros that could handle the ten numerical dig-its, all the letters of the alphabet, punctuation marks, and other symbols. After conversion to NMOS, the faster 8080 microprocessor, together with its competitor the Zilog Z80, was the heart of home-build computer kits such as the Altair, which were quickly bought for tinkering by teenage boys.

Among the older boys, Steve Wozniak built the first general-purpose, compact, stand-alone home computer in 1976, and Steve Jobs promoted the derivative Apple II so aggressively as to force a reluctant IBM to join the personal computer parade with its PC, an open system effectively mandated by fear of Antitrust investigation, and so run by CPUs from Intel, memory from TI, and an operating system from start-up Microsoft.

IBM's open system allowed cloning by Compaq and price reduction by new manufacturers such as Acer in Taiwan, who sold PCs all over the world under their own brand, but with lower production costs, came to OEM almost all the other PC brands, including IBM, Dell, HP, Toshiba, and Sony.[8]

While the capabilities of even super-thin notebook computers have far surpassed the bulky early acronymic mainframes, much greater computer speed and power are required for physics and chemistry calculations. quantum mechanics, nuclear fusion reactions, astrophysics, fluid dynamics, molecular structure, weather prediction, global warming, and all manner of simulations, and indeed later the training datasets and multiple parameters of artificial neural networks (ANNs). Internet and Cloud information are scoured by the fast-growing *large language models* (LLMs) and processed by *generative pre-trained transformers* (GPTs), and *Artificial General Intelligence* (AGI) likely will be capable of almost any and all human intellectual undertakings, and in humanoid robot form, physical endeavors as well.

NOTES

1. For the basic physics of semiconductors and their development, see the author's book, Chen, R. H. 2011, *Liquid Crystal Displays, Fundamental Physics and Technology*, Wiley. For a detailed technical description of the commercial development of transistors, see Burgess, M., 2011, *Early Semiconductor History of Texas Instruments*.
2. Men who are proud of their soldering skill could not hold a candle to those young women who deftly soldered thousands of components every day. The number of soldering joints was estimated at 500k.
3. Spaghetti and lasagna images are from en.wikimedia.org and used under Creative Commons.
4. The idea of an *integrated circuit* was first proposed by the British radar expert G.W.A. Dummer in 1952, combining wired-together transistors, resistors, and capacitors in a single semiconductor block, but he never fabricated a working device.
5. Kilby "flying wire" drawing from US Pat.No.3,138,743; patent figures are all in the public domain.
6. A *flip-flop* is a circuit that can store two independent states in memory; drawing from US Pat. No. 3,150,299. Kilby was awarded the Nobel Prize for Physics in 2000 for the integrated circuit, but Noyce had died in 1990, and thus was ineligible for the award. Kilby graciously acknowledged Noyce's contributions in his Nobel acceptance speech.
7. In n-type (negative) semiconductors, the majority atomic charge carriers are the negatively charged electrons, and in p-type (positive) semiconductors the majority carriers are positively charged holes. Since the number of defects per wafer is generally constant, larger wafers and more and smaller transistors per unit area of wafer (small design rules) plus greater fabrication

efficiency meant more defect-free transistors per wafer, and higher production yields. bringing down the price of each chip; roughly every two years doubling the number of transistors on a chip and halving the price (Moore's Law).

8. "OEM" stands for "Original Equipment Manufacturer", which means two different things in the West (the branded product) and in the East (the contracted actual manufacturer).

CHAPTER **3**

Software

I N 1804, JOSEPH MARIE Jacquard invented a weaving loom that used a deck of punch-hole cards through which movable rods would inter-weave different color threads for colorfully patterned cloth manufacture (Figure 3.1). Different sequences of cards would control the loom rods to produce different colors and designs; the *Jacquard loom* was thus the first machine to be *programmed*, first for weaving and its progeny later promoted to interlace numbers in scientific and engineering equations.[1]

The Babbage *Analytical Engine* conceived in 1834 was to be controlled by the principles of Jacquard's loom by Babbage's assistant, the lovely daughter of the great poet Lord Byron, who devised the first ordered sequences of cards for the Analytical Engine's calculations. The self-taught mathematician Ada Countess of Lovelace in 1834 thus was history's first computer programmer.[2]

The Countess would solve mathematical problems by simply changing the order of the punch cards, as she wrote in her operating instructions,[3]

We may say most aptly that the Analytical Engine weaves algebraical patterns just as the Jacquard loom weaves flowers and leaves.

The order of the punch-hole cards thus constituted *instruction sets*, which would be seen as written in a *machine language* that Jacquard's weaving loom could understand, and later to drive the computations of the Analytical Engine.

DOI: 10.1201/9781003463542-3

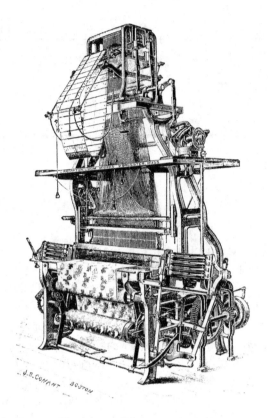

FIGURE 3.1   The Jacquard loom.

In a binary computer, encoded strings of zeros and ones constituted the instruction where, for example, in an 8-bit instruction, the first four bits may tell the computer what to do (command) and the last four bits tell it where the data to use can be found (address register).

There are more than 200 fundamental computer operations, so encoding and keeping track of all the instructions in machine language was a nightmare, made worse by the fact that different machines would have different designs and structures, and even if they were similar, there were no rules for program step sequences and register addresses, so a program written for one machine, even in fundamental machine language, was not transferable to another machine.

The computational logic problem had been solved by Turing and von Neumann, but the practical implementation of programming in machine language was difficult and error-prone, and the machine language instruction sets were almost impossible to decipher, even by the programmers themselves after the fact.

This provided the impetus for *software generalization*, and Grace Hopper, an assistant on the UNIVAC project, was assigned by John Mauchly to make their new machine, the *Binary Automatic Computer* (BINAC), capable of accepting equations nearly *as written* by the user.

Hopper realized at once that "one could use some kind of higher code other than the machine code" to program a computer, a code that was more easily understood by scientists and engineers that could be translated *by the computer* into machine language for the computer to translate and read.

Since many processes and mathematical computations are repeated during an operation, Hopper and co-workers devised short command mnemonics such as LOAD, GO TO, and SQRT which held such instruction sets in a subroutine library that could be simply called up by the program.

A sequence of such broader commands constituted an *assembly language* with the machine doing the work of translating from that assembly language to the detailed steps of machine language.[4]

Hopper in 1952 called the computer's language translation *compiling*, and the translator was called a *compiler*, terms that are still in use today.[5]

The next step was to formulate an even *higher-level language* for more diverse processing that used commands in more natural language and have the computer compile that language into assembly language and then into machine language.

Software development pioneer John Backus at IBM in 1957 created *FORTRAN* (Formula Translation) that could be compiled on the new IBM 704, going from high-level to assembly to machine language. FORTRAN convolved several assembly and machine language instructions into single statements that could command a computer to perform in a logical flow easily learned by scientists and engineers, for example, ASSIGN (say a variable), DO, IF-THEN, DO-LOOP, and ENDFILE and fundamental mathematical functions such as sin, cos, ln, exp, and so on, thus making computer operations so clear and logical that scientists and engineers could quickly learn FORTRAN by self-study.

When a computer is turned on, stored commands instruct the CPU to find the operating system (OS) which then takes control of the system hardware; a few functional programs are permanently stored in the ROM and the CPU executes the instructions in machine language after being translated from a high-level language computer program through an assembly language using a compiler; a RAM typically holds information and programs during processing.

Primarily for business and commercial use, the *Common Business Oriented Language* (COBOL) could more broadly file, sort, merge, add, subtract, and calculate percentages over large sets of data and conveniently generate business reports, practically in syntactical English. COBOL is still used by businesses and its older programs in the language are stored in many companies' mainframe legacy archives for reference.[6]

To spread more general use of computers, John Kemeny in 1964 wrote the simplified BASIC operating system and programming language, which is installed in almost every computer and is easily learned by almost anyone.

FORTRAN, COBOL, BASIC, and the European computer science-based *Algorithmic Language* (ALGOL) are based on the *imperative* programming approach that at any time the program has an implicit state defining the values of all the variables and current point of control, and as the program executes the von Neumann architecture, the programmer can know each and every step and state through which the computer passes by examining the program's *dump files*.

THE MINICOMPUTER

Beginning in the 1960s, computer time-sharing was centralized in a computer center that batch-processed computer punch cards. A user had to trek to the computer center, carefully carrying his box of punch cards lest they be dropped, ruining his program's sequence, and then after waiting in line to use the card reader, was fare-charged for his "computer time" computation, then wait for the printed output, and de-bug and amend on-site, again waiting in line to use the keypunch machine, put the new cards in order with the original program, and again wait for his batch to run, receive the paper output, and again check and correct, a process usually requiring a whole day or days and nights (the key being choosing the least crowded times, like late at night).[7]

This arduous and time-consuming centralized computing process was improved upon by dumb terminals that could input without punching cards and metamorphosed to "smart" distributed computing. The Digital Equipment Corporation's (DEC) series of newly coined *minicomputers* that they called *Programmed Data Processors* (PDP-n) did not have a mainframe's processing power, but they were smaller and cheaper and could be programmed and run on-site for specific tasks of the departments as well as linking to a mainframe.

At the time, however, each company's minicomputer came with its own specific design and operating system, so programs were written in machine or assembly language for each minicomputer, and if a company or department upgraded or changed to a different vendor, all the old programs and routines were useless and had to be re-written for use on the new minicomputer, and, of course, programs could not be shared with other companies or even departments in the same entity.

Ken Thompson at AT&T Bell Labs loaded his *Solar System Simulation* onto a PDP-7 with good graphics, but to his chagrin, it lacked basic *utility* routines such as file management and text editing, so he and his colleague Dennis Ritchie wrote the utilities in assembly language. They then loaded these routines into one of Bell Labs' new PDP-11s that was faster with more memory than the PDP-7.

Although impressed by Thompson's *Space Travel* graphics, many Bell Labs researchers were more interested in the utility routines for their own PDPs, and so in response to many requests, Thompson and Ritchie wrote up an *operating system* manual for the PDP-11s, calling the system *UNIX*, which was upgraded for time-share in 1972 and could hold 100 modules for recording instrument readings, sorting, linking, and analyzing. It could not only serve as a stand-alone multipurpose computer but manage program and data files, control laboratory equipment, text-edit, and data-format reports using a central *kernel*, and through a *shell*, time-share over multiple local and remote networks (Figure 3.2).[8]

Most of the scientists and engineers at Bell Labs, however, did not know assembly language, so Thompson and Ritchie wrote a higher-level language for programming the UNIX operating system (OS) that they called "*C*" (as the successor to the *Basic Combined Programming Language* called "*B*" then in use). Born in 1973, C is the imperative programming progenitor of the later-to-be declarable, object-oriented *C++* language now widely used for the coding of artificial intelligence algorithms.

The primarily mathematical logic and operational functions of imperative programming were expanded to *objects* in an *object-oriented* programming (OOP) approach that simulated systems by grouping data and instructions into direct and explicit representations of modular objects having a set of instructions and relevant discrete portions of data each representing one facet of a given OOP program.

Object-oriented languages such as *C++, PHP, Ruby, Java, Python*, and *JavaScript* would later be used for artificial intelligence programs and,

FIGURE 3.2   DEC PDP-11 minicomputer.

except maybe C++, are easier to learn and use; for example, JavaScript which is primarily used for Internet websites' dynamic interaction for notebook computers and smartphones.

The more basic or theoretical artificial intelligence algorithms typically use C or C++, which is closer to assembly language and is therefore more flexible from the ground up.

For programs written for a specific goal, *declarative* programming does not specify the language or method to be used but is given a *domain-specific language* (DSL), the objective, and perhaps a flowchart. The program (ideally) can determine the best way to achieve the goal and is thus an optimizing, an example is SQL.[9]

However, most artificial intelligence programs are written in imperative programming, but some are both imperative and declarative, depending on the task.

The popular imperative close-to-English syntax language Python uses an *interpreter* that allows code to be executed immediately after it is written, providing feedback for faster and more efficient programming. It has a large library of frameworks with algorithms for general programming and machine learning.

Because the *generative pre-trained transformer* (GPT) needs only simple instructions regarding what a program should do and can write the program in many different languages, it can be considered a declarative program.

ChatGPT employs the Python programming language, *PyTorch* framework, and *TensorFlow* platform, and mostly executes relevant data and information by *cut and paste* from a *large language model* (LLM), mainly the global network of computers *Internet* and infrastructure, platform, and software *Cloud Services.*[10]

These programs, tutorials, and programming techniques are all available for download or in host-computer coding platforms such as *GitHub* and *Red Hat* and cloud services for applications and websites like *AWS*.

Almost all basic computational tasks such as matrix algebra, calculus, differential equations, application programs, and so on are available from, for example, *Matlab, Wolfram*, and *Octave*, and specific engineering, physics, chemistry, and mathematics application and optimization software, as well as artificial intelligence algorithms are also available from websites and platforms on the Internet.

The platforms provide the *Linux* operating system with which one can design and build projects and provide large storage depositories, project management, version control, technical advice, bug fixes, blogs for user interactions and technical expertise for problem-solving.

Google's *tensor processing unit* chip (TPU) was announced in 2015, although not for sale itself, from 2018, its processing is available through the downloadable *end-to-end TensorFlow* platform that can be run on any 64-bit computer.

Between nodes in *TensorFlow's dataflow graph* is a tensor that relates the nodes, and in accord with the algorithm, it can rapidly parallel process the data, compute probabilities, and function relationships and interactions; that is, *TensorFlow* can efficiently process very large data sets with a huge number of weighting parameters for machine learning.

Almost all of the software is open source or available under Permissive Software Licenses (PSL), discussed in the next chapter.

NOTES

1. Figure 3.1 Jacquard loom figure (1891) uploaded by D.C. Llach is in the public domain.
2. The Countess was self-taught because women were not allowed to study mathematics in university; it was obviously the universities' great loss. Lovelace quote from *BBC Science Focus Magazine* (October 13, 2020). Babbage's Differential Analyzer was never built, but in 1991, when the

London Science Museum constructed a working machine using materials and techniques available at Babbage's time as proof of concept.

3. Lovelace quote from *BBC Science Focus Magazine* (October 13, 2020).

4. For example, to ADD *X* and *Y* in assembly language involves fetching the values of *X* and *Y* from their registers, thereby requiring knowing their addresses, commanding the ALU to add them together, and sending the result to another register address in memory.

5. During her work for the Navy on the successor to Harvard Mark I, Mark II mysteriously crashed, and upon investigation, Hopper found that a moth had wandered into the maze of circuitry and died in shock blocking an electric relay switch that caused the malfunction. And so the phrase "debugging the computer" became the common term for fixing any computer program malfunction.

6. An example of the syntactical nature and universality of COBOL, when Grace Hopper was stranded at a computer center in Japan and having difficulty in communicating with the non-English speaking Japanese programmers, she used COBOL commands, pointing to herself and writing "MOVE", and then pointing outside the center, then wrote "GOTO Osaka Hotel". "Hotel" in Japanese is just "hoteru". Rf. *Computer Languages* 1989, *Time-Life Books*.

7. Computer time charges unintentionally promoted efficient programming (mostly FORTRAN) to lower using the mainframe computer "costs".

8. Figure 3.2, MBlairMartin, copyright holder of this work, hereby publishes it under the <u>Creative Commons Attribution-Share Alike 4.0 International</u> license.

9. Critics of the arcane and hard to spot punctuation, parentheses, and brackets of *C++* can blame the *B* language for the overly concise memory-saving, assembly language-type commands to be used primarily by scientists and engineers who have an eye for detail. SQL, *Structured Query Language*, is a programming language for storing and processing information in a relational database.

10. All of these will be discussed in the following chapters.

CHAPTER 4

Open-Source Software

S OPHISTICATED SOFTWARE WOULD BE of little help in developing new
ideas and innovative algorithms if their source code and techniques
were proprietary and kept confidential. The free and open publication and
use of *open-source software* (OSS) for software development, particularly
using free mathematics, engineering, application programs, and platforms
on the Internet and in the Cloud, together with free host-computer coding platforms, all were critical to the development of artificial intelligence.

The open-source ethos was born in MIT's *Tech Model Railroad Club*
where a successful design of some complex model railway operation was
completed by someone who just "hacked away" at the circuits until the
model train behaved as planned. The good circuits of course were incor-
porated into the grand model and their designs were freely accessible not
only to all club members but to anyone interested, and in those days model
trains were very popular for boys of all ages.

Corporate open source, however, was incongruously hatched at AT&T, a
fierce protector of its intellectual property rights. The reason is that before
the 1960s, AT&T was an officially sanctioned monopoly of America's
telephone system under a Consent Decree by the Department of Justice
ostensibly to ensure compatible public utility services. Indeed, for generations, the stodgy black dial phone was the only telephone model available,
and all the switchboards, relay stations, and landlines were produced by
AT&T's manufacturing subsidiary Western Electric.

To prevent further expansion of AT&T's monopoly, the Justice
Department prohibited AT&T from any activities outside of

DOI: 10.1201/9781003463542-4

telephone communications, particularly noting the burgeoning new field of computers.

So computer companies did not have to contend with AT&T, and particularly IBM's 700 series and then the 360/system series mainframes were dominating the industry (and raising antitrust investigations themselves).

Harnessed by the DOJ Consent Decree, AT&T's UNIX for inter-computer communication was distributed free to universities under a *General Public License* (GPL), and for a nominal $20,000 for commercial use. The publication of *C* in 1974 on DEC's $50,000 PDP-11, running on the UNIX OS, soon became the standard inter-computer link for science and engineering.

In 1977, the *Berkeley Software Distribution* (BSD) operating system based on Unix and developed at UC Berkeley was licensed and widely used in academia and well-known workstation companies like Sun Micro. The derivative *FreeBSD* is now used by, among others, Netflix, AWS, and Cisco as an OS platform.

Berkeley established a free and open host-computer coding platform that was the harbinger of today's platforms such as GitHub and Red Hat that are critical for the development of artificial intelligence.

Since any computer with a UNIX OS and a *C* compiler could employ the UNIX or BSD system, by 1983 more than 80% of university computer science departments had adopted UNIX and BSD inter-computer communications systems.

However, in the same year, the Department of Justice's antitrust consent agreement was lifted and after the government-mandated dissolution of AT&T into regional "Baby Bells" (and the grievous disbanding of the great Bell Labs), UNIX suddenly became a commercial product in high demand with a high price; only the early source code was licensed in object code and its copyright vigorously asserted.[1]

And so, although creating the concept of a General Public License, AT&T returned to its commercial roots and the specter of computer software copyright was used for establishing an onerous royalty-bearing, secret source code licensing regime, subsequently adopted by, among others, Microsoft, and the infringement suits that would be aggressively pursued by Apple.[2]

Meanwhile, back at MIT, the *Tech Model Railroad Club*'s system was becoming too complicated for circuit boards, so the Club requested permission to time-share on MIT's new IBM 704. But they were rebuked with the harsh words, "the 704 has more important uses", and TMRC was reduced to begging for computer time on the PDPs in various MIT departments.

However, DEC came to their rescue in 1970 by generously donating two brand new PDP-11s, and while still serving to control the model railway, the free-spirited young men informally promulgated a completely unrestricted flow of ideas while they had fun developing new computer games.

The hacker spirit quickly crossed the country to Palo Alto's *Homebrew Computer Club*, and the magazine *Popular Electronics* spread the Hacker Gospel to all points in between, where boys with a mathematical/electronics bent could try their hand at the new tech, so that not only Apple but also Microsoft and many other pioneering companies were eventually spawned in the juvenile but heady atmosphere of computer hardware and software development, spurred on by the free and open exchange of ideas and programs.

The most popular home-build kit was the Altair 8800, which was featured on the January 1975 cover of *Popular Electronics*. It ran on an Intel 8080 CPU and the DOS assembly language operating system (Figure 4.1).

While still a high school student in Seattle, Bill Gates spent most of his time playing with his Altair 8800, but he crossed the hacker Rubicon when he lambasted the Homebrew boys for copying his Altair BASIC program, which he felt was worth $500.

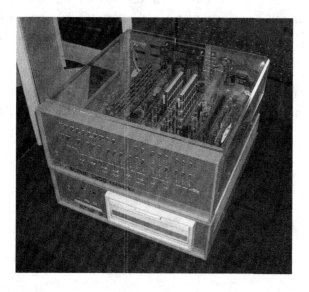

FIGURE 4.1 Altair home-build microprocessor.

In 1976, enthusiastic Homebrew member Steve Wozniak paired his Altair 8800 with a TV screen and a keyboard, and together with an operating system and application programs, he produced the first personal computer, Apple I (Figure 4.2), and true to the hacker spirit, freely distributed both his hardware and software designs. But his eyes were opened to the commercial world with the next event in his life.[3]

One day, while Wozniak was as usual reading various electronics journals at the Stanford Linear Accelerator library, he noted an article about AT&T's primitive 2600 Hz audible switching tones that were used for the very expensive long-distance calling of the time. He electronically mimicked AT&T's audible tone for long-distance calls and demonstrated his *Blue Box* to his 17-year-old friend Steve Jobs, who, still in high school, displayed the first indications of his soon-to-become famous marketing ingenuity.

He reasoned that university students would make many long-distance calls home to former high school classmates and friends at different schools, so he went to nearby Stanford and Berkeley dormitories and marched down the halls knocking on doors to demonstrate the Blue Box. He quickly sold more than 100 Blue Boxes for $170 each, earning substantial pocket change for himself and Woz.

But it was more than the money, Jobs credits their Blue Box for an epiphany,[4]

> *If we wouldn't have made blue boxes, there would have been no Apple. Because we would have not had not only confidence that we could build something and make it work ... but we also had the sense of magic that we could influence the world.*

However, at that time, the two Steves were only changing the long-distance telephone call world; AT&T's cash cow and soon the police and even the

FIGURE 4.2 Apple I personal computer.

FBI were closing in on Jobs and Woz, and they abandoned their venture, but the commercial promise of new technology captivated Jobs, or was it to be the other way around ... ?

Steve Jobs wanted to establish an enduring company that would use new technology for exquisitely designed tech-fashion products that provided new utility that would *influence the world*. Steve Wozniak, however, was by nature, a man who wanted to share his ideas, with little regard to fame and wealth.

The decidedly less charismatic but opportunistic Bill Gates was fortuitously riding on the IBM PC and clones' operating system monopoly to make exorbitant rent-seeking profits. The never-ending new *Windows* versions reflect Gates' concentration on a commercial strategy of changes in old versions that almost forced consumers to install (not necessarily useful or better) new *Windows* versions.[5]

Apple and Microsoft, although conceived in the hacker spirit, in effect replaced AT&T and IBM's anti-competitive behavior with their own and were busily restraining the trade for personal computer operating systems and applications software with onerous licensing programs (Microsoft) and vigorous enforcement of patents and software copyright (Apple).

Jobs was known to often visit the Stanford Research Institute (now SRI) and the Palo Alto Research Center (PARC) to see the new cursor controller mouse and overlapping windows. Microsoft for its part copied the Seattle Computer Company's CP/M for its later MS-DOS. BSD was sued by AT&T for copyright infringement of UNIX in 1992, and notwithstanding the severe intellectual property enforcement zeal of AT&T, Microsoft, and Apple, Microsoft later hypocritically used BSD-produced code in Windows 2000, and Apple's macOS and IOS operating systems were largely based on BSD's progeny *FreeBSD*. It seems that in those days of the new personal computer industry, one can have one's cake and eat it too.[6]

UNIX, Apple, and Microsoft's closed systems rankled the young members of the fast-growing academic discipline of computer science. One, the acknowledged computer science genius Richard Stallman, took it upon himself to develop a competing inter-computer operating system he called "GNU", a recursive acronym, "GNU's Not UNIX", the source code of which was public, free, and distributable and included a toolkit to encourage further development of GNU and applications.

Mounted on the open-range wildebeest GNU, trampling through the halls of computer science departments, the wild-eyed Stallman offered open-source software for all to freely use and distribute, all the while excoriating Apple and Microsoft's denials of public access to source code as veritable "crimes against humanity" that stifled the growth of the new computer software discipline and the widespread use and development of new computer programs.

In his fight against secret and proprietary software, Stallman's crusade found a kindred spirit during a speech in far-away Finland. Inspired by his talk, the University of Helsinki graduate student Linus Torvalds, using the GNU toolkit, developed *Linux* as an open-source competing operating system for personal computers. He published the source code online in 1991 under a GPL, attracting many young hackers and inviting them to improve it and develop new utilities and application programs in an early instance of crowd sourcing.

Apple, however, has retained the closed-system, overpriced hardware and software with Jobs' now *de rigueur* new product stage performances that propelled the Mac, iPod, iPad, and iPhone to become the darlings of the high-tech *fashionista*.[7]

Stallman still fought on though, and in 1985 he founded the Free Software Foundation (FSF) to promote completely open-source software and denounced any form of restricted access, including Digital Rights Management (DRM), software patents, and restrictive software licensing. He called his creed "copyleft" as opposed to "copyright" and asserted that "copyright infringement is not a crime, hindering the right to freely use software is the crime".

Stallman invoked the "Free and Open Source Software" (FOSS) movement, where "free" meant no charge and no bounds and "open" meant disclosed and available, all based on the plaintive desire of "information longing to be free".

He was resolute, for while the world mourned the death of the tech icon Jobs in 2011, when asked for his reaction, ever-true to his open-source creed, Stallman said, "I'm not happy that he died, but I am glad that he is gone".[8]

Not unexpectedly, software companies viewed Stallman and Torvald's ideas and actions as simplistic idealism, lacking an understanding of business. Investments must have a return, and if there are no returns then there is no investment and no new software development. FOSS was nothing more than unrealistic fantasizing that would only rouse the copyists to appropriate innovative research and development by others. The software companies insisted on proprietary rights, particularly in downstream software transactions.[9]

It can be understood that the notoriety that motivates a boy to freely show off his coding skills may not suffice for the man who has a family to feed. The creativity and hard work in his field no doubt deserve proportionate remuneration, but, on the other hand, the free and open exchange of information and ideas are necessary to its development, as traditionally in scientific research, and now critically so for the science and technology of artificial intelligence.

Both sides had reasonable concerns and a compromise was necessary, the first being the *Open Source Initiative* (OSI), which allowed the commercialization of software but also mandated the publication of source code.

This obviously was not acceptable to the software companies as profitable software would be immediately copied.

So the tech idealists at the University of California, in conformance with the Berkeley quintessence, together with kindred spirits at the citadel of enlightened engineering MIT, devised a GPL derivative, the *Permissive Software License* (PSL), and established the *BSD* and *MIT License* regimes that advocated the open-source development of new software, but with a nod to entrepreneurship, required only that the source of the open-source software must be attributed and critically did not require publication of the source code, and companies retained intellectual property rights in downstream transactions of their products.[10]

Thus, a potentially lucrative software product could be protected as intellectual property, rewarding its creator with exclusive rights, but the basic mathematics of algorithms and their processing was open source. In this way, commercialization and possible profits could encourage new ideas, but the mathematical tools necessary for manifesting the idea would be free and open.

Whether PSL will succeed in satisfying the concerns of both sides remains to be seen, but there is no doubt that PSL will instigate lots of business for law firms.

The major PSLs are from the MIT License and the BSD, and the early users were Mozilla, Apache, and the Cluster Computing Framework.

Almost all the basic software for artificial intelligence development is free and available online, for example, C++, *Python*, *Java*, and *JavaScript*, and linear algebra and other mathematics programs from, among others, *Matlab* and *Octave*, and free host-computer coding platforms *GitHub* and *Red Hat* offer massive parallel processing by, for example, *TensorFlow*, and for-pay extra services, such as larger depositories, product management, version control, consulting, extra security, and so on. But software companies are still applying for patents and registering copyrights, just in case intellectual property wars break out.

As for the property rights of software developed at the host-computer coding platforms, the users are the owners of the software they develop there, but the platform generally has a right to sub-license with terms negotiated case-by-case, but, for example, *GitHub* will almost always receive some handling charges.

The fact that *GitHub* and *Red Hat* had never made a profit with their business model was not a deterrent to Microsoft who acquired *GitHub* in 2018 for $7 billion, and the erstwhile symbol of strictly proprietary technology opened up and published its hitherto secret source code and brought arch-rival Linux into its global cloud Azure, and thus by acquisition adopted the new tech ethos.

The next year IBM acquired *Red Hat* for $34 billion, particularly for *OpenShift*, the user-spaced *containers* for hybrid cloud application software development.

However, since the courts have declared databases copyrightable, artificial intelligence datasets will likely become proprietary, and with algorithms mostly open source, AI research has turned to the quality, as well as quantity, and control of data. For example, avoiding false, contrived, and possibly unethical or non-politically correct data, although such filtering may produce inaccurate results.

It will be interesting to see how stodgy proprietary property companies like Microsoft and IBM will adapt to the youth-driven tech of the times, and if *GitHub* and *Red Hat* can remain independent of their giant owners going forward.

It has come to pass that the conflict between the harsh commercial despotism of AT&T, Apple's Steve Jobs and Microsoft's Bill Gates against the dreamy idealism of Richard Stallman and Linus Torvalds would find common ground in a compromise between crass rent-seeking and idealistic free and open development of the software of artificial intelligence.

NOTES

1. The copyrighting of source code as literary expression is by itself debatable because the program's object code of sequences of 0's and 1's is hardly "expression" protected by copyright, rendering such protection a product of commercial interests rather than legal theory. Furthermore, a work protected by copyright should be fully disclosed and published for society to

appreciate and license. Proprietary software source code has long been an exception as being neither published nor licensable (the object code is licensed for use), a legal anomaly exploited by Microsoft and Apple. The author while at Acer Computers negotiated many agreements with Microsoft and Apple and experienced first-hand their vigorous assertion of intellectual property rights.

2. Software patents have a long and varied history, but because of the limitation of patent scope generally to the machine using the software, the principal protection of software has been copyright.

3. Altair 8800 image from Michael Holley's 2004 photograph, dedicated to the public domain, Apple I image from free use license of Wikimedia Commons.

4. Haden, J. Aug. 1, 2019, Inc.

5. "Rent-seeking" has been defined as "increasing one's share of existing wealth without creating new wealth". Gates had piggybacked on IBM's personal computer and disproportionately increased his share of the wealth based on the debatable extension of copyright laws to source code and object code. In terms of economic analysis, with virtually no capital expenditure and practically zero reproduction costs, Microsoft collected exorbitant licensing fees and copyright infringement awards out of proportion to its minimal creation of new wealth. This economic waste of licensing fees, court awards, and attorneys' fees could be better spent in R&D or for charitable causes (which to his credit, Gates did after retirement). Microsoft's secret source code also inhibited research as evidenced by the fact that practically all new software research is based on the GPL-licensed GNU, Linux, and FreeBSD operating systems; Microsoft finally came around after Gates had left and opened up its source code for public use. Apple with its innovative and sleek products has created wealth, and deserves profits, but overly so by exploiting low-cost off-shore supply chains, planned obsolescence, and extreme overpricing of its severely closed-technology products.

6. The lawsuit brought by AT&T's Unix System Laboratory was justifiably settled in BSD's favor.

7. Aside from its designed-in incompatibility with other brands, for example, the iPod's non-replaceable battery and iPhone6's deliberate slow-down and accelerated battery run-down to goose sales of later (more expensive) models are instances of Apple's commercial subterfuges.

8. Quote from Stallman's personal blog.

9. Stallman once suggested that the federal government could apply an excise tax on software that would be used as royalties to software developers.

10. GitHub was acquired by Microsoft in 2018, the speech recognition company Nuance in 2021. IBM acquired Red Hat in 2019 to operate out of its Hybrid Cloud Division, but it remains to be seen whether the commercial interests override the open software ethos (an example is the 2023 firing and re-hiring after four days of (non-profit) OpenAI's boss Sam Altman, the controversy being true to the "Open" part of AI and greater commercialization despite fears of ChatGPT-6 being smarter than humans.

CHAPTER 5

Expert Systems

I N 1956, AT THE inaugural *Dartmouth Summer Research Project on Artificial Intelligence*, the AI pioneers John McCarthy and Marvin Minsky set out their goal of a pure *top-down* artificial intelligence machine in a grant proposal,[1]

> *The study is to proceed on the basis of the conjecture that every aspect of learning or any other feature of intelligence can in principle be so precisely described that a machine can be made to simulate it.*

The conjecture was philosophically and theologically based on Descartes' *mind-body problem.* Spinoza believed in *monism*, that an abstract thought and physical matter are not separable; therefore, a thought could be produced by a piece of physical matter. However, Descartes' *dualism* meant that the mind and matter are completely separate, and therefore matter by itself could not have a mind and could not have an abstract thought; that is, thinking matter, a thinking machine, was not possible.[2]

Two attendees at the conference, Allen Newell and Herbert Simon, apparently Spinoza believers, picked a task that they believed would be an ultimate proof of the conjecture: a machine that could prove mathematical theorems.[3]

From the rules of symbolic logic as set forth in 1910 (and later editions) by Alfred North Whitehead and Bertrand Russell in their epochal three-volume tome *Principia Mathematica*, symbolic logic is used to prove mathematical theorems. Logic symbols were used to represent *tautologies*

DOI: 10.1201/9781003463542-5

such as "All unmarried men are bachelors", *material implication* (also called *material condition*) $a \rightarrow b$; *transitive property* (also called *inferences*) such as "if A implies B ($a \rightarrow b$), and B implies C ($b \rightarrow c$), then A implies C" ($a \rightarrow c$), *syllogisms* of two propositions resulting in a deduction, such as "all cats are mammals, all mammals have four legs, therefor all cats have four legs" (a deduction that may be invalid because of a false proposition), and the mathematical concepts of *uniqueness* ($\exists!$) and *completeness* (for every $\Gamma \vDash \varphi$ it is also true that $\Gamma \vdash \varphi$).

The *Principia* was no model of getting to the point quickly, but the epitome of rigor, beginning from first principles and famously taking 29 logic equations to prove that 1 is a number and 379 pages to prove $1 + 1 = 2$.

The Logic Theorist, conceived by Newell and Simon and programmed by Cliff Shaw in 1956, used the innovative *Information Processing Language* (IPL), a forerunner of the *List Processor* (LISP) artificial intelligence language, converted words and phrases into binary-coded symbols, and by comparing and matching symbol strings in a tree search where the root is the hypothesis, each branch is a deduction founded on symbolic logic, and the objective theorem is at a twig of the tree; the proof is the trajectory from root through branch to twig.

There are many different distinct choices and varied combinations that one can take along the way to reach the goal, just as there are many ways of reaching a solution of most mathematical equations. One can try all the possibilities one by one, checking the result, and if wanting, backtrack to try the possibility of a different branch path in the tree. Even for restricted domain problems (such as mathematical theorems and board games), this *exhaustive search* method, however, would encounter a *combinatorial explosion* of different possible combinations of possible paths.

McCarthy and Minsky's conjecture of a "top-down" approach, such as that of the Logical Theorist, came to be called an *expert system* that employed hardwired microprocessors programmed for logical operations as a *means-end* analysis informally called *hill-climbing*, that is, because a mountain climbers step up rather than down is generally more likely to contribute to the goal of reaching the summit.

However, reflecting the complexity of any endeavor, a step down might lead to an unforeseen lower-level roped passage to the top, so the hill-climbing trajectory approach is obviously not absolute; examples in mathematics might be proof by contradiction (*tertium non datur*), and in games, *queen sacrifice* in chess and *tenuki* and *sente* in Go.[4]

Nevertheless, the Logic Theorist actually succeeded in proving 38 of the 52 theorems in Chapter 2 of the *Principia*, in many ways different from the proofs by the authors themselves but to the delight of Bertrand Russell himself. But if Russell's *logicism* is the touchstone of the *Logic Theorist*; that is, basic axioms are thrown into a logic machine, and after undergoing logical processing, outcome proofs of mathematical theorems (!), then mathematics depends entirely on the operations of pure logic to derive all of the fundamental theorems, a belief that also was espoused by great mathematicians such as Russell and David Hilbert; their mathematics philosophy that came to be called *formalism*.

But the pure logic of formalism can lead to contradictions; to take a simple example, the statement "I am lying", which if true is not true because you are lying, and thus it appeared that logic alone could not by itself drive a thinking machine.

With mathematics ability as the test of pure intelligence, the ancient philosopher Plato and the father of modern science Galileo saw mathematics ability as a gift from Heaven bestowed to a select few, and the eminent mathematicians Brouwer and Poincaré believed mathematics was based on *intuition*, also bestowed from Heaven.

Top-down expert systems typically break a big problem into smaller sub-problems, tackling each in turn by *if-then* program steps and integrating all the sub-problems into a machine process that in principle could solve any problem.

For example, if the problem is to win an American football match, each series of downs is a sub-problem, a traditionally trained head coach would reason: *if* you are in your end of the field and *if* it is fourth down and 7 yards to go, *then* punt. However, the score, game strategy, and personnel might dictate otherwise; for instance, you have a good fake-punt play, a punter who can run and pass, and a coach whose perceived conservatism would result in surprise. The expert system could objectively take into consideration other options, not the least because it can disregard unsubstantiated coaching homilies and accept the assumption of risk (short-time losing the game and long-term getting fired).

Success in tackling the sub-problems integrated into the total match will likely result in victory, so a football-coaching expert system, through thorough searches of possibilities and probabilities resulting in logic-derived actions, should perform better than a human head coach.

THE FIFTH GENERATION

The height of optimism for top-down expert systems was on display in Japan in the early 1980s. It was a time of an ascendant Japan threatening to dominate the world economy with its semiconductors, consumer products, and automobiles. The Ministry of International Trade and Industry (MITI), in an attempt to administer Japan's *coup de grâce* for advanced technology supremacy, announced a ten-year project to develop thinking machines for commerce and industry and further to converse, translate, and logically reason for societal activities.

This *Fifth Generation* of computing would have powers of reasoning based on symbolic inference systems connected to a central knowledge database and distributed by inter-connected base machine systems.[5]

The big idea was that natural resources-poor Japan could thrive on the exploitation of its formidable human resources and, with the aid of expert systems, solve the problems of limited resources, environmental damage, aging populations, education, and language differences and promote more efficient production and communication.

The culmination of the plan would not only enhance manufacturing capability and quality by employing expert tooling machines but fully computerize research and development to propel Japan to an elite position as the foremost knowledge-based economy in the world. Furthermore, as an exporting economy, Japan could export *information*, a resource that did not deplete (like oil) but rather continuously *increase*, providing Japan a continually growing new export commodity.[6]

Alarmed, the United States also did some planning for their own Fifth Generation, but after a few starts and stops, development never got off the ground. And just as well for America and unfortunately for Japan, the technology, computer programs, and data were not sufficient at that time for such an immense undertaking, and after ten years, the project was abandoned in 1990 having achieved none of the announced goals, and in the reverse of the hope for AI technology, cynicism of artificial intelligence in general brought on the second "AI Winter" of frozen funding for expert systems and later, freezing active artificial intelligence research for the next ten years.

NOTES

1. *Proceedings of the Dartmouth Summer Research Project on Artificial Intelligence* 1956.

2. *Monism* implies that upon the material destruction of the brain, any consciousness or ideas are also destroyed, and death means the end of any type of "being" contrary to many religious beliefs. *Dualism* means that mind and matter are distinct, and therefore some kind of conscious "being" may exist after death, perhaps the "soul" or "reincarnation" of many religions.

3. Herbert Simon won the Nobel Prize in Economics in 1978 for his work in organizational behavior.

4. Proof by contradiction is assuming a proposition to be false and showing that this leads to contradiction; tried and true, it was formalized in the *Principle of the Excluded Middle* (PEM), but criticized by Dutch mathematician L.E.J. Brouwer for not considering a third "undecided" or "half-right" proposition instead of the rigid "there is no third option". See Hallman, H. 2006, *Great Feuds in Mathematics*, Wiley. *Tenuki* means ignoring an opponent's moves, accepting a possible local loss to gain a later initiative (*sente*).

5. Part of Japan's enthusiasm expert systems to translate and converse likely came from the immense differences between Japanese and Western languages, causing the Japanese no little angst, and here perhaps contributing to an overly hopeful belief in the utility of expert systems.

6. FeigenbaumE.A and P. McCorduck 1984, *The Fifth Generation*, Signet. The first four computer generations are designated (1) vacuum tube, (2) transistor, (3) integrated circuits, and (4) very-large-scale integration (VLSI).

CHAPTER 6

Inverted Decision Trees

IBM's FIRST COMMERCIAL, GENERAL-USE computer, the IBM 701, as programmed by Arthur Samuel, beat a checkers champion. The natural next step was for an IBM computer to play chess, a far more complex game, and to many, a chess champion is the epitome of supreme human intelligence.

After 30 years of development of microprocessors, integrated circuits, and programming, computer scientists believed that a championship-level chess expert system was a possibility.

A chess board has 64 squares and 32 chess pieces with different move rules; all the possible combinations of both players' positions are almost infinite at 10^{120}. So a chess-playing expert system would need a more sophisticated version of an inverted tree search algorithm than Samuel's checkers-playing tree search.

Although the origin of the tree search is cloudy, it is believed that the *inverted decision tree* was first used in 1959 by England's William Belson, reasonably enough, for biological classification; the first inverted decision tree computer program was developed at Berkeley and Stanford in 1972 and later refined, notably by Claude Shannon, and then used in artificial intelligence algorithms.

The inverted decision tree with a *root node* at the top and branches descending from the root node, with each branch having *children* branches leading to other nodes. The *branching factor b* is the number of children connected to each node, and the *depth* of the tree *d* is the number of *nodes* of a decision-making path from the root down to the farthest leaf node; for

DOI: 10.1201/9781003463542-6

example, the total number of moves made in a chess game. The inverted decision tree for a depth of 3 is shown schematically in Figure 6.1.

The *depth* or *level* of the decision tree designates the number of layers of branch nodes counted from the top (tree root), or in other words, the depth of a search for an optimal move, and for accomplished players, roughly a measure of how far ahead the player can plan his moves ("lookahead"). A level of seven layers makes a mediocre player and about 15 levels are required for a Grandmaster.

It can be seen from the figure that there are many more nodes as the decision tree depth increases, in fact, they increase exponentially, in this case meaning the number of nodes y doubles as the depth increases, $y = 2^d$, where d is the depth, but the number of possible combinations of those nodes is the exponent of an exponential increase; that is, an exponential increase of an exponentially increasing number, which can grow very rapidly. For example, the factorial $y = d!$ is the increase of an increasing number, and its increase for say $d = 20$ compared to a simple exponential is

$$y = 2^{20} = 1,048,576 \approx 1 \times 10^6$$

$$y = 20! = 2,432,902,008,176,640,000 \approx 2.4 \times 10^{18}$$

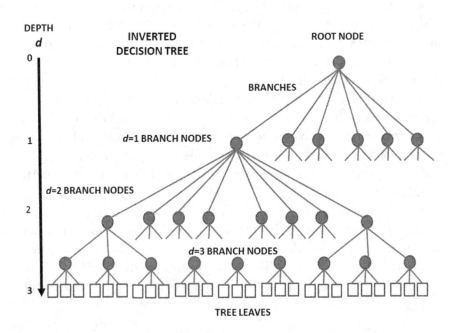

FIGURE 6.1   Inverted decision tree.

As can be seen, the factorial, as d increases, is a considerably higher number, and the cause of the phenomenon of *combinatorial explosion*. In a typical chess game, a depth of $d = 100$ (50 moves per player), and b varies with the game and stage of the game; so if the average is $b = 10$, the number of leaf nodes (*complexity*) is given by $b^d = 10^{100}$, which means that for an expert-level player traversing every combination of branches in the whole decision tree, as estimated by Shannon, will amount to about 10^{120} possible combinations of node traverses that can be made in a game.

This number is more than the total number of atoms in the Universe, so going down every branch in *exhaustive searches* is not feasible, even for a very fast computer; that is, the number of nodes and path combinations must be reduced while retaining the accuracy of the tree search.[1]

MINIMAX

A *minimax* algorithm from game theory attempts to find the optimum move under the circumstances of the board at hand. The minimax algorithm is based on the fact that players always want to optimize the utility of their moves by rationally maximizing the value of their own move and minimizing the value of the opponent's moves, so each player is a maximizer (MAX) of their own moves and a minimizer (MIN) of the opponent's moves.

After the opening book (a series of traditional opening moves such as the Sicilian Defense, Ruy Lopez, and Queen's Gambit and many variations of the same) decided upon by a chess computer's professional consultants and stored in an opening book RAM, there are no evaluation numbers at each tree node except at the leaf nodes at the bottom of the (inverted) search tree, as shown in the simple decision tree of depth = 2, nodes (x and y) at depth = 1, and *branching ratio* $b = 2$ (number of branching emanating from a node) (Figure 6.2).

As the game progresses, the leaf node evaluations S are calculated by a *linear scoring polynomial g* that describes the board's *features f_i* at the time, such as positions of the knights, pawn structure, castling possibility, rook files, and so on,

$$S = g(f_1 + f_2 + f_3 + \ldots + f_n)$$

Then multiplying every f_i by a constant c_i to assess (weight) the value of each feature in the particular chess piece layout at hand, S is analyzed and determined by consulting professional chess master,

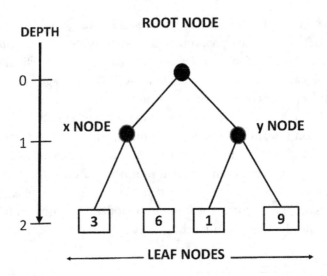

FIGURE 6.2   Search tree depth = 2.

$$S = g(c_1 f_1 + c_2 f_2 + c_3 f_3 + \cdots + c_n f_n)$$

The idea now is to find the path to the desired leaf node using the minimax algorithm, either MAX for offense and MIIN for defense (minimize the opponent's MAX), and the move depending on the current look-ahead possibilities based on strategy, tactics, and the opponent's moves.

Starting from the leaf values as a base, the players will climb "up" the decision tree by evaluating the node values that are initially based on the value of the desired leaf node, engage in the minimax procedure, comparing branch node values until they reach the *static decision root node* that determines the move. It is assumed that each player will rationally select nodes based on minimax; one player, the expert system, bases its moves on the decision tree algorithm; the game being adversarial, the opponent should act rationally in accord with minimizing or maximizing.

An example of the minimax procedure for a simple decision tree of depth 2 is shown in Figure 6.3.

After the openings, say the opponent's first move MAX(0) is to the x node, which is now the present state of the board; the expert system "sees" from the x node branches to the leaf nodes, and if it wants to minimize MIN(1) the opponent's MAX(0), there is a choice of 3 or 6 and will take the 3 branch because 3 < 6 will minimize the value of the opponent's MAX move, so node $x = 3$ (the expert system could of course instead try to maximize the efficacy of his move at x and take the 6 leaf node, but assume

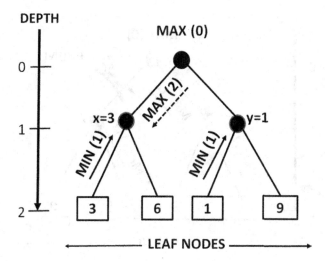

DEPTH

MAX (0)

0

1

x=3   MAX (2)   y=1

MIN (1)   MIN (1)

2

| 3 | | 6 | | 1 | | 9 |

← ———————— **LEAF NODES** ————————→

FIGURE 6.3   Minimax decision tree.

here that minimizing the opponent's MAX move is tactically preferable). If the expert system takes the y node instead, the branches lead to 1 and 9, and because the lower value MIN(1) of the leaf nodes is 1 (1 < 9), so it chooses node $y = 1$.

In light of the relatively low value of the x node, it would be preferable for the opponent to maximize his next move; the rational choice is between $x = 3$ and $y = 1$; the opponent MAX chooses the branch leading to the $x = 3$ node as well since 3 > 1.

That is how the nodes are evaluated and how the players choose the nodes; the level of choices has been moved up one level, stepwise raising the level of the move decision-making higher up the decision tree, finally to the static decision root node.[2]

α–β PRUNING

Now it can be seen in this case that neither player has chosen the y node in this minimax round, so it can be *excised*; that is, all the branches and nodes emanating from y can be ignored, and the search reduced without having any following effect on the minimax process and the results of the game itself, as shown in Figure. 6.4.

This is called α–β *pruning* because there are two parameters (α and β) to compare. How much can α–β pruning reduce the search computation? Professional players look ahead an average of ten moves and the number of nodes has been calculated in that case after α–β pruning is

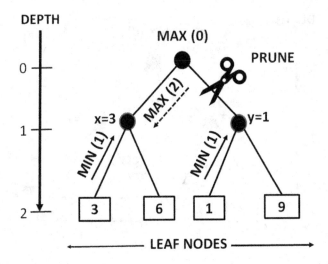

FIGURE 6.4   Decision tree α–β pruning.

reduced from b^x to $2b^{x/2}$ which, because it is exponential, is a very significant reduction.[3]

PROGRESSIVE DEEPENING

Another method of reducing the size of the search, yet ensuring that an
optimal move is found at the fewest number of levels, *Iterative Deepening
Depth-First Search* (IDDFS), thankfully also called *progressive deepening*,
which provides an assurance that if the minimax process is stopped (perhaps because some desired node has just been found), whenever and wher-
ever in the game, the time, complexity, and storage will all be reduced and
the best node values will have been determined according to the minimax
algorithm to that depth.

In *Depth-First Search* (DFS), a recursive search of all nodes to the depth
of a desired node (meaning one that has a particularly optimal evaluation
for the particular situation) is not found on the current path to a final
childless node; it will backtrack on the same path to find a branch with
nodes and go down through those branches, repeating in this way along
all the different paths while labeling the nodes visited and storing in a
stack so to avoid visiting a node more than once (and avoiding a possible
infinite loop in the program).

Iterative Deepening (ID) of the Depth-First Search "controls" DFS by
restricting it to a given depth; however, if the desired node is not found at
that depth, the depth is increased and the search iterated until a satisfactory

node is found; this at least ensures that the best node value will be found under the circumstances. The search then can be terminated at that depth and there is no need to go deeper, and the amount of minimax computation, time, and complexity are reduced.

The chess expert system using minimax and progressive deepening is now ready to play chess.

NOTES

1. 10^{100} is called a "googol", the root of the name "Google" the tech company.
2. Doing the α-β pruning algorithm is easier than explaining it, and there are canned programs, Abbeel, P., UC Berkeley, "Machine Vision", Spring 2013 online, and Sebastian Lague, "Algorithms explained – minimax and alpha-beta pruning", YouTube April 20, 2018. Minimax is further described in Chapter 18, Game Theory. For minimax theory related to chess, refer to Winston, P., "Minimax", MIT Fall 2010 online lectures on artificial intelligence.
3. RefWinston, P., 2010, Fall, "Minimax", in *Lectures on Artificial Intelligence*, MIT.

CHAPTER 7

Deep Blue

AFTER THE NEWS BROKE out that Arthur Samuel's IBM 701 had defeated the national fourth-ranked and Connecticut champion checkers player Robert Nealy, IBM's stock price rose more than 15%, and the erstwhile recalcitrant Thomas Watson Sr. made a U-turn and his son Thomas Watson Jr. launched a series of one million-dollar prizes in what he called *Man versus Machine Grand Challenges*.

Possibly at odds with humanity's long-term self-interest, fair matches of supreme mental combat in the arenas of two of the primary indicia of human intelligence, IBM's *Grand Challenges* in Western chess and Google's *DeepMind* foray into the ancient Eastern game of *Go*, the old and new tech companies were both exploring the potential of the new artificial intelligence.

With IBM 701's victory as backdrop, the natural next step was to extend the machines' ken from simple checkers to sophisticated chess, believed by many to be the ultimate test of human intelligence in the Western world; this time, engendering fear among the populace was not a concern, seeking wonder and subsequent publicity and income for IBM was the goal.

IBM 701 had long retired, and the step-up in complexity to chess was formidable, but in the 30 years since 701's victory, the processor chips, integrated circuits, and software advances made IBM confident that a chess-playing computer could compete at the very highest level.

However, IBM had no such machine in development. But, at Carnegie-Mellon University, an electrical engineering graduate student from the National Taiwan University, F.H. Hsu, was working on the hardware for just such a machine, and IBM duly recruited him and his entire team plus

DOI: 10.1201/9781003463542-7

a cohort of consulting professional chess players and (critically) provided a generous budget.

Through the 1980s, Hsu and his team developed *Deep Thought*, a chess-playing machine that performed the *inverted decision tree search* using hardwired *move generators* to process a minimax algorithm with α–β pruning and progressive deepening. The leaf node values were determined by a linear combination of weighted features evaluated by professional chess consultants and published records of Grandmaster matches. Opening books and endgames were provided by advisers and stored for random access.

Now the question was whether a computer could master the intricacies of a complex game like chess. An experiment with champion chess players as subjects found that they could almost immediately memorize and recognize *rational* board layouts but could not recognize irrational layouts, so it was concluded that professional chess players play by the pattern recognition of the formation of the pieces on the board and, through memory, innate talent, experience, heuristics, and skill, can quickly and rationally respond to a logical board.

The chess professionals who can play dozens of players at once and *speed chess* (where each player has only 30 minutes for the entire game) are demonstrations of their supreme pattern recognition capability.

A face screwed up in confusion during a match is a sure sign of an irrational board disposition contrary to their experience and logic. A chess computer is electronically performing pattern recognition from expert consultants' leaf node evaluations based on the weighted features $c_i f_i$ of the board layout; that is, $S = g(c_1 f_1 + c_2 f_2 + c_3 f_{3 + \ldots + } c_n f_n)$ but perhaps may be discombobulated by an irrational move or weird board pattern, just like human chess professionals may be.

For example, when Hsu's Deep Thought, the first computer to defeat a Grandmaster in a sanctioned game, was experimentally presented with the admittedly not-likely-to-encounter hypothetical board layout shown in Figure. 7.1, it would be clear to a human player that black has an enormous advantage in pieces with two rooks and a bishop, while white has only a king and pawns.[1]

A draw could have been easily achieved by just moving the white king around behind the line of pawns. But because it was available for material gain with no immediate personal threat, Deep Thought logically took the black rook with its pawn, and in doing so destroyed the pawn's line of defense, resulting in an inevitable defeat.

FIGURE 7.1   Pawn-line defense.

Even a pedestrian human player, having seen the pawn pattern on the board, would not have made such a mistake, but Deep Thought apparently did not recognize the unusual board pattern and was top-down programmed to capture pieces and attempt to checkmate black's king. Deep Thought did what it did, and the expert system in this case was no better than a bungling amateur, proving that although "expert" it could also be stupid.

Deep Thought's successor Deep Blue in 1993 with more memory, faster processors, and more chess-playing accelerators should be able to avoid Deep Thought's naïveté, and after fully ten years of work designing the hardware drivers (Figure 7.2) and inverted decision tree algorithms (Figure 7.3), IBM believed that Deep Blue was ready to challenge Garry Kasparov, the then reigning world champion and legendary chess genius to an officially sanctioned traditional six-game championship challenge match.[2]

Kasparov at just 17 gained Grandmaster status, and at only 22, he was the youngest-ever world champion, a position he held for a record 255 months; he was generally recognized as the greatest player in the history of the game.[3]

Like all great champions, Kasparov assiduously studied his opponent before the match to find playing proclivities and weaknesses. From his pre-match training with other chess computers, Kasparov evidently realized that because of the sheer thoroughness of a high probability search

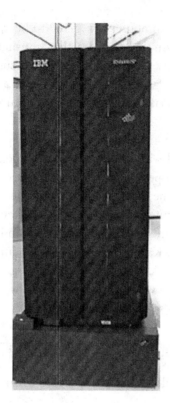

FIGURE 7.2 Deep Blue.

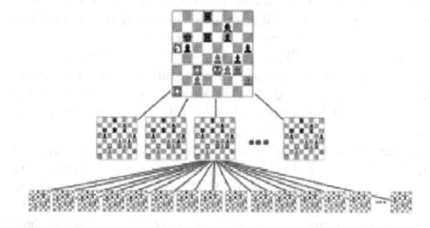

FIGURE 7.3 Deep Blue search tree schematic.

for good moves that a computer can bring to bear, a straight-up match would be difficult to win, but the idea that unusual moves or irrational patterns could confuse a computer germinated, such that Kasparov for one, believed that this was the key to victory over a chess computer; that is, deliberately suboptimal moves are made to confuse the rationally wired computer.

So Kasparov devised an "anti-computer" strategy of deliberately playing strange moves and weird board layouts; he believed that Deep Blue would not be able to properly respond, just like Deep Thought encountering the strange pawn-line defense, he believed that responding to "weirdness" was a chess computer's weakness that could be exploited.

In 1996, Deep Blue lined up against Kasparov in Philadelphia for a six-game challenge match. Kasparov, employing his anti-computer strategy, handily won the first match in Philadelphia 4-2, but the next year in New York City in May 1997, a world champion match was officially sanctioned by the professional chess governing body, the *Fédération Internationale des Échecs* (FIDE).

This time, arrayed against Kasparov, was an upgraded Deep Blue with 36 newly designed chess accelerator processors and 316 special chips running a massively parallel RS/6000 SP Super Workstation chess-playing computer.[4]

Aware of Deep Thought's vulnerability, Deep Blue employed a decision tree that could handle extraordinary moves and unusual board patterns with "blow-up searches" that could deviate from the routine minimax evaluator operation of the inverted decision tree.

Deep Blue's specifically designed high-performance processors and re-designed software could minimax tree search 50 billion possible positions at an astounding rate of 200 million moves per second. After α–β pruning and progressive deepening of the decision tree, a tree depth of six to eight levels was searched to evaluate optimum moves.

An example of an evaluator function is the well-known tactic that rooks and bishops could be advantageously placed on or near files where there is an option of early "opening up the files" by removing the pawns on the files (typically by pawn exchange) and thereby exerting long-range and long-term pressure on the opponent's heavy pieces. The file of course must not be opened prematurely allowing the opponent to notice and disrupt or challenge. The rooks and bishops rather should be set up on a "potentially open" file and the player should wait for an opportune moment to open the file.

The value of a decision tree node S is calculated by functions g which are linear combinations of f_i chess piece positional features, weighted by a factor c_i that reflects the immediate situation of the pieces on the board,

$$S = g(c_1 f_1 + c_2 f_2 + c_3 f_3 + \ldots + c_n f_n)$$

The match attracted worldwide attention and in the audience were the usual Grandmasters and professional players, together with chess critics, news reporters, the general chess public, and a good number of Kasparov fans but, atypically, also a considerable contingent of computer scientists and CS students.

In Game 1 of the Match even with white advantage, Kasparov abandoned normal openings, and his heavy pieces never left his own half of the board even when Deep Blue's black bishop potential diagonal was evident as shown in Figure. 7.4.

For the first nine moves, the pieces were closed and highly positional, but in a disconcerting *10.e3* move, Kasparov confirmed his suboptimal laying-back strategy, as a normal move would be a *10.e4* offering a pawn exchange.

Deep Blue, however, did not capitalize on this suboptimal move to gain the initiative because, in this instance, its automatic tuning of the evaluation function had decreased the weighting for certain types of moves, but

FIGURE 7.4   Deep Blue vs. Kasparov Game 1.

in extreme cases, the minimum weight was reached, and this saturation meant that Deep Blue no longer distinguished a very bad position from an even worse position. Kasparov believed his strategy was working.

The saturation bug was found and corrected after Game 4, but after a few more questionable moves, Deep Blue lost Game 1, but knew why it lost, and Kasparov only gained an expected white victory in the opening skirmish. But in winning, his belief in his suboptimal move lay-back strategy was reinforced.

In Game 2, Kasparov's anti-computer lay-back strategy fell apart when Deep Blue, affirmatively responding to the lay-back, pinned down Kasparov's heavy pieces to his back rank utilizing rook and bishop potentially open files and diagonals, taking the offensive and a final victory.

In light of his success in Game 1, Kasparov did not believe that Deep Blue could adapt and counter his strategy without human operators recognizing the suboptimal lay-back positioning and unexpected moves. He complained after the match,

> You know, [the anti-computer strategy] was working, but suddenly it stopped working – suddenly Deep Blue found a way just to break the pawn chains and start a confrontation in a very, very convenient situation.

Deep Blue's team assiduously denied what both sides had agreed upon about non-intervening during a Game. After Game 2, Deep Blue's computation log was checked by some chess commentators and it was believed that Deep Blue's upgraded software could adjust to Kasparov's strange moves.

Deep Blue's last move in Game 2, *Ra6*, deemed a "highly dubious but excellent move" by commentators, was sufficiently shocking to end the game with a full point for Deep Blue. Kasparov's suspicions and accusation of cheating no doubt affected his playing and were likely factors in his missing a substantive chance for a draw in Game 2, as pointed out in after-game analysis by his seconds, and his black thus had missed a chance for a critical half-point.

After two days' rest, Game 3 began with Kasparov, even with a white advantage, opened with *1.d3*, a move that likely had never been played before in a championship match, and waited for a weird response. But Deep Blue instead took advantage and set up a pawn-chain defense (Figure 7.5). With both sides playing defensively, a draw could be predicted, and

FIGURE 7.5   Deep Blue vs. Kasparov Game 3.

Kasparov again lost the white advantage and half a point. At the end, the lonely little *1.d3* pawn was still there forlornly waiting for a chance to prove his mettle.

In Game 4, after the strange moves in Game 1 attributed to evaluation software bugs were repaired, upon Kasparov's 42nd move, Deep Blue *self-terminated* owing to a piece of code that monitored the efficiency of a parallel search and terminated the program if the efficiency dropped below a given level, which it did, causing Deep Blue to shut down, much to the operator Hsu's chagrin. Furthermore, according to the human vs. computer chess match protocol, the time to re-boot was charged to the computer side.

With less time, Deep Blue's computations were time limited such that deep searches were not possible, and with Kasparov still making anti-computer moves, Game 4 was stultifying, and at Deep Blue's 56th move, Deep Blue's *endings ROM* indicated a rook ending draw that Kasparov also saw, and playing black, he expediently offered, gaining a half-point.

Game 5, after three days' rest, started out with Deep Blue, although playing black, aggressively pushing his heavy pieces out, and had a significant advantage in piece development, while white again strangely remained back in its own end (Figure 7.6). Deep Blue then suddenly sent an *h*-file pawn down in a disconcerting *11.h5* move, startling all the commentators and even the Deep Blue consultants.

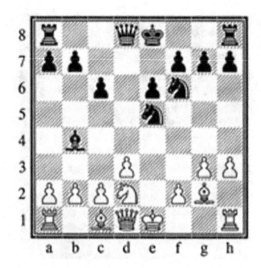

FIGURE 7.6   Deep Blue vs. Kasparov Game 5.

However, the move meant to Hsu that Deep Blue was warning the greatest player who ever lived, "if you castle kingside, I will attack you!" It was an audacious threat that no human player would ever dare level at Kasparov.

Indeed, Kasparov post-match lamented, "no computer plays *h5*!" But Hsu knew exactly what Deep Blue's hardware was doing,[5]

> *When I saw the move **h5** from Deep Blue, I knew precisely what hardware evaluation features prompted the move. During the last two months of chip design before the rematch, I added drastic changes to the hardware for king safety evaluation. Before the king castles, the hardware computes three sets of king safety evaluations, one for kingside castling, one for queenside castling, and one for staying in the center. The real king safety evaluation is the weighted linear combination of the three, with the weighting based on the relative ranking of the three, and difficulty of making the castling moves. ... In the game position, Deep Blue could always castle queenside safely, and therefore move **h5** was perfectly capable from its point of view.*

Although Kasparov later had a passed pawn ready to promote, Deep Blue just marched its king forward and initiated a drawing sequence based on repetition checks, so Game 5 was drawn, with Kasparov's white once again only salvaging a half-point.

After five games, the score was tied 2½–2½ and for the final Game 6, unless Kasparov playing black could defeat a Deep Blue with a white advantage, history would be made with a computer tying or defeating a reigning world champion in a FIDE-sanctioned championship match.

Before Game 6, the commentators now almost all believed that once Deep Blue had the initiative, it could not be stopped. Indeed, Deep Blue was out of its opening book with *11.Bf4* as shown in Figure. 7.7, and seeing three pawns worth of positional compensation, it was clearly in attack mode. Spurning material gains for an ultimate king kill, Deep Blue continued its attack and at *19.c4*, Kasparov resigned. History was made with the final board shown in Figure 7.8.[6]

Despite a clear loss in Game 6 and the Match, a combative Kasparov refused to acknowledge Deep Blue's superiority, still believing that the IBM team had cheated. He had previously demanded to see Deep Blue's game logs during the Match, but IBM refused on the entirely reasonable grounds that that would be tantamount to revealing game strategy while the game was ongoing, akin to a human telling his opponent his strategy and tactics during a match. IBM did agree to provide complete game logs after the Match to show that there was no in-game human intervention.

Deep Blue's game logs did reveal a random error in Game 1, and commentators speculated that Kasparov interpreted the subsequent fixes instead as Game 2 in-game changes by the Deep Blue team; in other words, Kasparov would not accept that he could be beaten by a machine.

FIGURE 7.7   Deep Blue vs. Kasparov Game 6.

FIGURE 7.8   Deep Blue vs. Kasparov Final.

The closeness of the match was not definitive of the superiority of machine over man, but Kasparov's paranoia throughout, and his abysmal performance in Game 6 could at least establish that Deep Blue's ineluctably cold logic could triumph over the warm frailty of human emotion and the pride of extremely self-aware human beings.

A rematch was discussed, but Kasparov's demands of further bizarre perquisites for himself and his team, and his proposed three-week match with two- and three-day rest periods (obviously for Kasparov to physically and mentally recoup, Deep Blue of course needed no rest periods).[7]

Sought-after sponsors felt that such a rematch format was too long to hold public attention and were not forthcoming. IBM, having put up the $700,000/$300,000 winner/loser prize for the Match, paying for the considerable expenses of New York's poshest skyscrapers hotels, and accommodating Kasparov and his entourage every amenity, having already achieved its goal, was not enthusiastic.

Deep Blue's victory over Kasparov demonstrated that a top-down expert system could defeat a formidable human in a very restricted domain, but the in-between games hardware and software fixes revealed flaws in the "expert" system.

Inevitably, if computers are the best players, won't human vs. human matches lose their appeal, and tarnish a Grandmaster's cerebral distinction, for the real champions are computers?

At least there was still some public interest in a match between the new world chess champion Norway's Magnus Carlsen and American challenger Fabiano Caruana in the 2018 World Chess Championship. It, however, ended in 12 draws, ultimately being decided by a penalty kick-like rapid chess confrontation (total 30 minutes per player) won by Carlsen.

Carlsen and other GMs nowadays do not play championship matches against computers but rather use them to hone their skills. The acclaim that accorded a human "World Chess Champion" no doubt will be eroded by chess computers, and perhaps even worse, because of similar training on them, the computers may be indirectly the reason for the many stultifying draws of high-level human matches.

After its stunning victory, could Deep Blue, like humans, be *self-aware* of its superiority? It will never be known because the fate of most innovative research devices is *dissection*; the RS/6000 SP was sent back to IBM's test floor and the shell returned, but two cards went to IBM headquarters in Armonk for visitor demonstrations, and the rest were inserted into the older version RS/6000 SP and dispersed to various workstations and spare parts shelves.

In the press conference after the match, Kasparov was cheered and heartily encouraged by an audience including many chess masters, expert commentators, the press, and the general public, but when IBM's Deep Blue team assembled on the stage, their notable technical achievement notwithstanding, they were met with thinly veiled disdainful murmuring.

There is no sin in standing up for humankind against a machine, but Deep Blue's victory evoked unease, fear, and even hostility, and the audience apparently sensed menace rather than hope. Perhaps Watson Sr. was right after all.

NOTES

1. Deep Thought defeated the "Great Dane" Grandmaster Bent Larsen in 1988. The "pawn-line defense" episode and figure were described in J. Seymour & D. Norwood 1993, *New Scientist*, 139, No. 1889, 23–26.
2. Deep Blue image by Jim Gardner who dedicated it to the Public Domain. Inverted tree search schematic figure from Stanford University CS221 course, private communication.
3. At just 22, Kasparov was the youngest world champion, and the reigning champion for an unprecedented 225/228 months from 1986 to 2005, his 1999, FIDE Elo ranking of 2,851 was the highest in history until Magnus Carlsen, a Grandmaster at 13, scored 2,882 in 2014, the highest ranking to date.

4. Thin 30 node, 120 MHz P2SC microprocessors at each node and 480 specifically designed VLSI chess accelerator chip cards running C language software under the AIX operating system on a 32-bit microchannel bus.
5. Chess board layout figures from F.H. Hsu 2002, *Behind Deep Blue: Building the Computer That Defeated the Chess World*, Princeton University Press under license from Creative Commons.
6. After writing a book about his match with Deep Blue, Kasparov dialed back his pique saying, "I am not writing any love letters to IBM, but my respect for the Deep Blue team went up, and my opinion of my own play, and Deep Blue's play, went down. Today you can buy a chess engine for your laptop that will beat Deep Blue quite easily". He did not add that by logical inference, that chess engine could also easily defeat him as well.
7. Psychiatrists and social psychologists literally had a field day with Kasparov's well-documented prima donna persona; he and his extremely protective mother Clara demanded all manner of amenities, including a private dressing room with designated furniture, specially prepared exotic snacks and drinks, bathroom facilities that only he could use, his very expensive match chair, a custom-designed (by Garry) chess clock made by Audemars Piguet, special player timing controls giving the human an advantage, and so on and on. All that might be laid by the psychologists at the door of a quintessential Mama's boy deserving and always getting the best.

CHAPTER 8

Jeopardy and Miss Debater

D EEP BLUE'S VICTORY OVER Kasparov, although eye-catching, was in the very *closed domain* of championship chess. In 2011, IBM Watson Grand Challenged two all-time champions in the popular "given the Answer, pose the Question", a completely *open-domain* television game called *Jeopardy* where virtually any subject could come up for the three contestants to vie to first respond correctly.

Since the other two contestants could not have any reference information in hand, relying only on their immediate memory, to be a fair contest, IBM Watson could not access external information, say from the Internet, so its memory and processors had to be completely self-contained.

Thus, backstage in a supposedly sound-proof holding room, there still could be heard the muffled sound of the large fans cooling ten refrigerator-sized Power 750 Servers running Acer eDC Framework Services that stored 200 million pages of information for 3000-core processor massively parallel processing that could be called a *heterogeneous ensemble of expert systems.*

Because *Jeopardy* is a contest of information knowledge and retrieval and not a test of speech processing, the moderator's oral "Answers" were simultaneously given to Watson in text to optically character recognize (OCR), and after search and retrieval, the "Question" was text-to-speech (TTS) synthesized.

IBM Watson parsed and parallel processed information threads based on keyword matching, factoids, grammar, and verbal relationships in a

DOI: 10.1201/9781003463542-8

knowledge graph, which is a graphic database of objects, events, situations, and concepts, and the relationships among them. It could find and relate information in an average of 3 seconds, while also performing risk management of responses (there is a penalty for incorrect responses and contestants can wager higher winnings in the later stages of the contest). The parallel-processing IBM Watson could quickly jump from one to another of the connections of the graph nodes to show how the combinations were playing out to determine candidate "Questions".

Because it is open domain, the "Answers" could be about almost everything, so the contestant who answers correctly can choose a category from a list, for example, world history, geography, countries, science, literature, sports, movies, TV, music, pop culture, food and drink, politics, and so on. The one-sentence "Answers" were very specific, for instance, "First planted in Rio in the 1770s, by 1830 it was Brazil's top export", and the correct Question is "What is coffee?"

Accurate text recognition and TTS within the seconds range that champion contestants take to press the buzzer to pre-empt competitors required sophisticated textual natural language processing and extremely high search and relational processing speeds.

Watson used WordNet, Wikipedia, and many other information sources and was reinforcement learning-trained against a hundred previous *Jeopardy* winners. In the "championship" contest, Watson soundly defeated legendary *Jeopardy* champions Brad Rutter and Ken Jennings and won a million-dollar Grand Challenge prize and a further $77,147 in game winnings. It could have continued to win every contest it entered but understandably was banned from further *Jeopardy* competition. After the defeat, a chastened Jennings displayed an equanimity that Kasparov sorely lacked but added a warning for the future,[1]

> *Just as factory jobs were eliminated in the 20th Century by new assembly-line robots, Brad and I were the first knowledge-industry workers put out of work by the new generation of "thinking" machines. "Quiz show contestant" may be the first job made redundant by Watson, but I'm sure it won't be the last.*

MISS DEBATER

Jeopardy was an open-domain contest of information knowledge with no element of active disagreement. How about the very open domain of discussions in everyday life; that is, the very human bane of arguing?

There is no subject or opinion under the Sun (and beyond) that cannot be argued, and everyone and their spouses are "experts" in any subject. The opinions in such raucous disagreements, however, generally are not particularly accurate, sometimes emotional, and mostly illogical, with scant persuasiveness that seldom produces a winner.

However, arguments can also be formalized in a decorous debate, which requires accurate knowledge of almost any subject, rapid information processing, creative construction of an argument, persuasive exposition of a position, quick understanding of the opposition's argument, and then analytical deconstruction and sharp rebuttal of the opponent's points, all buttressed by references and data. To win, a debater must convince a skeptical audience of a superior point of view, first by cogent information and logic and finally by persuasive elocution, often including emotive and even unexpected humor.

Each side has 15 minutes to prepare, 7 minutes to present their argument, 4 minutes for rebuttal, and 2 minutes for a closing argument.

The first debating IBM Grand Challenge was held in San Francisco in • 2019, the proposition was: "Should the government subsidize space exploration?" IBM's Project Debater argued, supported by facts, that space exploration benefits humankind because it can help to advance scientific discoveries and inspire young people to think of greater things beyond their immediate environment.

Project Debater was developed at IBM in Israel, and Noa Ovadia, the 2016 Israeli national debate champion, in opposition, argued that there are better applications for government subsidies such as directly for scientific research here on Earth. Project Debater rebutted with historical facts that much government subsidy spending often leads nowhere, whereas the potential technological benefits from space exploration have produced, for example, communication satellite rockets, the use of computers, and the invention of non-stick *Teflon* for frying pans. Noa countered by eliciting the high cost of space exploration, the uncertainty of any useful results, and the more direct benefits to society of government-sponsored research.

In their allotted times, both sides indeed presented succinct and emotive arguments, and in accord with the debating rules, a pre-debate poll *for* and *against* the proposition is held and results are recorded; the debater who produces the greater number of cross-over votes after the debate is the winner. Project Debater won by changing the minds of a greater number of the audience.

In a second debate, the proposition was, "We should increase the use of telemedicine". Project Debater once again won a greater number of cross-over votes. Faced with a previously unknown subject, proposition, and no *a priori* choice of side, with only 15 minutes to research the proposition, in the far-reaching, diverse, and dynamic environment of an open-domain professional debate, the machine had once again won.

The next year, Project Debater was improved and now named *Miss Debater* in respect of its synthesized female voice, probably in an attempt to assuage the natural fear of robots, boldly took on the 31-year-old world champion Debating Officer of the Cambridge Union Society, Harish Natarajan. The proposition was: "Should the government provide pre-school subsidies?" Miss Debater greeted her opponent respectfully but tinged with a veiled menace,[2]

> *I have heard that you hold the world record*
> *in debate competition wins against humans,*
> *but I suspect that you have never debated a machine before.*
> *Welcome to the future!*

Harish hesitated but courteously nodded in response and the great debate began.

Miss Debater's reflexively gleaming façade's three blue orbs tantaliz-ingly rotated as she searched her *knowledge graphs* database of over ten bil-lion sentences from 300 hundred million newspapers, scientific journals, and other articles and then meticulously but quickly formed an argument.

Harish was also combing his memory, but instead of computer memory fetches, and microprocessors churning logic for arguments, he was busily scrawling notes on a notepad outlining arguments for his position.

Miss Debater produced academic research data and media quotes to show that subsidizing preschools is not just a matter of finance but a moral and political duty to protect some of society's most vulnerable children; again, a seemingly evocative argument rising above practical matters, albeit unfortunately sounding a little like a politician's speech.

Sensing and taking advantage of a general disdain for politicians among the public, Natarajan countered that too often subsidies function as politi-cally motivated giveaways to the middle class and that even with subsidies there will be many who still cannot afford pre-school for their children, a realistic but rather cynical rebuke in response to Miss Debater's moralistic point.

Although Miss Debater occasionally flashed un-machine-like humor during the debate, she was likely incapable of cynicism, and perhaps because of the rather harsh political climate of conservative self-help of the times, Harish the human won over more of the audience to his side.[3]

After the debate, a commentator noted that "While Miss Debater was clearly better than Harish at culling information and citing relevant facts, Harish had better rhetoric and better counters, and brought up tough rebuttals that Miss Debater did not adequately address".[4]

After the debate, Natarajan, a man from the East but brought up in the West's debating tradition, proposed an amalgam of East and West for the future development of artificial intelligence,[5]

> *After the first minute of getting used to the shock of it not being a human being, or the surprise of what that actually meant, it became much like a human being ... what the machine is better at than any human could ever be is finding relevant evidence, studies, examples, cases, and get that context. Another thing I think is very impressive was not only [Miss Debater's] ability to present evidence, but also its ability to explain why it matters in the context of the debate. Combining Project Debater's skills with those of a human would be incredibly powerful!*

Is Man still in control and the Machine as helpmate the best combination? Miss Debater's demonstrated fact-finding prowess, logical argument formation, debating ability, and persuasiveness, for better or worse, sooner or later may very well obsolesce the entire class of lawyers in our society ... and perhaps more distressingly, dispose of human and human teachers and professors as well.

NOTES

1. Quoted from J. Ford 2011, *Paging Dr. Watson ...*, singularityhub.com.
2. Reference for further details, see IBM research director Arvind Krishna's online blog.
3. For a description of the debate, refer to Fortune, February 12, 2019, and many other online accounts.
4. Commentator's quote from Inevitablehuman.com, February 12, 2021.
5. Natarajan's quote from *Project Debater*, top500.org.

CHAPTER 9

The Perceptron

M ORE THAN 2,000 YEARS ago, in his famous *Theory of Forms*, Plato asserted that ultimate reality exists beyond the physical world; the father of modern physics, Galileo, would say that one of the ultimate *Forms* is mathematics, afloat in the Heavens and bestowed by God to only a select few humans.

CYBERNETICS

One of those humans was Norbert Wiener, who graduated from university in mathematics at the age of 14, further studied zoology and philosophy, could speak seven languages (and was said to be difficult to understand in all of them), wrote his Harvard dissertation on the mathematical logic of set theory, and received his Ph.D. at the tender age of 17.

He was soon a member of the mathematics elite, traveling to Cambridge to learn from the legendary philosopher/mathematician Bertrand Russell and the renowned pure mathematician G.H. Hardy, and thence to the European citadel of mathematics and physics to study with the great David Hilbert at Gottingen.

With this transcendent résumé, the eclectic Wiener first taught philosophy at Harvard, but then following continually diverging interests, he took jobs as an engineer at General Electric and, of all things, a reporter for the *Boston Herald*.

With America's entry into World War I in 1917, eager to serve, but failing enlistment because of poor eyesight, Wiener was invited by the venerable mathematician Oswald Veblen, then serving as a Captain in the

DOI: 10.1201/9781003463542-9

army, to the Aberdeen Proving Ground to work on artillery shell ballistics, something that all governments bade their best mathematicians to do in wartime.

After the Great War and being rejected for a permanent position at Harvard, for which he (and Albert Einstein) blamed the anti-Semitic views of Professor G.D. Birkhoff, Wiener took a job as an instructor in mathematics at MIT.[1]

In the yeasty environment of Cambridge, Wiener joined regular meetings with intellectuals from many different disciplines, and with his own varied background, he began an independent study of the boundary between different disciplines, not today's interdisciplinary studies, but the *nexus* of zoology, neuroscience, medicine, mathematics, and electronics, the new computers, and finally philosophical ruminations about the mind-body problem.

Only 20 years after World War I ended, World War II began, and Wiener returned to his ballistics experience to approach the vexing problem of hitting high-flying and fast-moving enemy aircraft with slow-to-respond anti-aircraft guns. The targets were always moving, and for fighters, not necessarily in predictable patterns, and different weather conditions, changing winds, and the proclivities of the anti-aircraft guns themselves, all amounted to a very complicated problem with which the manual-mechanical anti-aircraft firing directors of the day could not cope.[2]

Here Wiener's studies in zoology and the many discussions with zoologists at the universities in Cambridge led him to the idea of the *nexus* between all creatures, including humans, to their environment, something called the *adaptive feedback loop* by which information is relayed from the environment to the creature, who adjusts with appropriate responses. As Charles Darwin wrote in his *Origin of the Species,*

> *It is not the strongest of the species that survive*
> *nor the most intelligent*
> *but the one most responsive to change*

It followed that a machine (like an anti-aircraft gun) must adapt to the environment of enemy aircraft, and so Wiener created the discipline of *cybernetics* ("steersman"), with the implication that a machine could adapt, and a veiled implication, that it could also "think".[3]

PENETRATING THE FOG OF WAR

Meanwhile, the war was going badly for the Allies. After the evacuation from Dunkirk and the surrender of the Low Countries and France, an air attack proposed by Hermann Göring was set to cripple Britain's naval and air defenses, followed by a blockade and Hitler's *Operation Sea Lion* cross-Channel invasion, altogether designed to force Britain to sue for peace, freeing the Nazis to turn East in their pursuit of *lebensraum*.

The campaign against Britain began in the Summer of 1940 when hundreds of Heinkel, Dornier, and Junkers heavy bombers and Ju-87 Stuka dive bombers pounded Britain's ports, shipping centers, airfields, and infrastructure.

When the bombers arrived with Messerschmitt fighter cover in daylight, Britain relied on human lookouts and telephones for communication; the early warning allowed the Hawker Hurricane and Spitfire fighters time to courageously rise to meet the enemy in the air. Many of the bombers were brought down by the Hurricanes who were in turn preyed upon by the Messerschmitts with whom the Spitfires fought in an air combat of relentless and horrific attrition.

Half of the defending pilots and aircrew, some 520 men, were killed in the air battle. Why were they alone in the defense? Where were the anti-aircraft guns? Indeed, there were 264 anti-aircraft guns with the number doubling after two days, but they could not hit the enemy aircraft and indeed it was Churchill's "Few" to whom "so much [was] owed" who saved the day in the First Battle of Britain.

On October 14, 1940, 380 Heinkel and Junkers bombers arrived over London, and although 8,326 anti-aircraft rounds were fired, the AA guns shot down only two of the slow-moving heavy bombers flying in formation.

Fully aware of the aiming problems, anti-aircraft gun-laying was often relegated to blanket firing to an altitude in an area in front of where the bombers were believed to be proceeding, hoping that they would simply fly into the hail of exploding proximity-fuse shells and destroy themselves.

Needless to say, such wishful tactics could not stem the tide of bombing; it was imperative that the tracking of the bombers and the aiming of the AA guns improve.

In the succeeding nighttime bomber raids of the Blitz that terrorized London well into 1941, the Hurricanes and Spitfires could not see the enemy to engage them in air battle. Britain's air defenses thereupon fell entirely upon the anti-aircraft guns, and although floodlights and fixed-baseline acoustic locators could spot the formations of approaching bombers, because of the shortcomings of the anti-aircraft aiming systems, London at night was virtually defenseless,[4]

> We had depended on anti-aircraft guns ... and apart from a solitary salvo loosed at the beginning of the raids, no gun had been shot in our defence ... we felt like sitting ducks.

Britain's best minds were brought to bear on the problem at the Royal Antiaircraft Command under the direction of the distinguished physicist, P.M.S. Blackett, and included the well-known mathematical physicists Ralph H. Fowler, Douglas Hartree, and Edward A. Milne.

Scientific anti-aircraft targeting begins with the mathematical physics of these two exemplary non-linear ballistic differential equations for the trajectories of projectiles fired from AA guns in the xy-plane as functions of the time t,

$$\frac{d^2x}{dt^2} = -\frac{C_d A \rho}{2m}\left[\left(\frac{dx}{dt}\right)^2 + \left(\frac{dy}{dt}\right)^2\right]^{1/2}\frac{dx}{dt}$$

$$\frac{d^2y}{dt^2} = -\frac{g}{m} - \frac{C_d A \rho}{2m}\left[\left(\frac{dx}{dt}\right)^2 + \left(\frac{dy}{dt}\right)^2\right]^{1/2}\frac{dy}{dt}$$

where $g = 9.8$ m/s^2 is the gravitational acceleration, m is the mass of the projectile, ρ is the density of the air, and C_d is the drag coefficient, which depends on the geometry of the projectile, and $A = (\pi d^2)/4$ is the frontal area of the projectile.

Non-linear differential equations cannot be solved in closed form, so they were set up numerically like Babbage's *Difference Engine*, and young women were recruited to compute the *ballistic firing tables*, increment by increment, using adding machines.[5]

However, the more than 750 different multiplications for each trajectory with 2,000 trajectories per calculation were something that even extremely diligent humans could not accurately perform by hand, keeping in mind that any errors could have devastating consequences. Fortunately, Vannevar Bush's *Differential Analyzer* machines (also operated by young women) could more quickly and easily produce the theoretical trajectories of the anti-aircraft shells.

The artillery shell trajectory mathematics was sound and the solutions true, with implementation helped along by no little anti-aircraft gun-laying heuristics, but their initial reckoning depended on manually operated optical trackers that supplied target range and bearing values from ballistic firing tables that were consulted to mechanically turn the shafts and gears of the *fire directors* of the anti-aircraft guns, setting elevation, range, and direction.

By the time the gun was ready to fire, however, the targets had gone on, conditions had changed, and the whole targeting procedure had to be repeated, often to little or no avail.

Meanwhile the Nazis were preparing the fast V-1 and the supersonic V-2 rockets ("flying bombs") attacks that would be even more devastating than the bombers; the situation was dire.

First to the rescue was the newly developed *radar* (to which Arthur Samuel contributed while at Bell Labs) that could provide the real-time day and night continuous position, speed, and direction tracking of aircraft by displaying a moving blip on an oscilloscope screen. This tracking dot was to be the scourge of enemy aircraft from this time forward, but more accurate gun-laying still had to be able to follow the blips to shoot down the bombers and missiles that showed up on a radar oscilloscope; that is, the anti-aircraft guns needed the enemy planes' position and flight trajectory feedback.

Twenty-nine-year-old David Parkinson at Bell Labs had been working on automatic *level recorders* that measured and controlled voltages to provide stable and uninterrupted voice communication in AT&T's telephone transmission lines.

A potentiometer responding to voltage changes controlled a pen recorder writing on a moving strip of paper, and Parkinson, apocryphally inspired by a dream, realized that in reverse the pen recording could just as well cause the potentiometer to produce an analogous electric voltage that tracked the motion of the pen, and so the motion of a radar blip on an oscilloscope screen could produce a potentiometer voltage that followed the blip and thus control the fire director of an anti-aircraft gun in a *continuous feedback loop*.

In the Winter of 1942, Bell Labs delivered such an electronic analog fire director to the army; the *M-9 Predictor*, which tracked the radar blip and electromechanically directed massive 90-mm breech anti-aircraft guns to shoot down invading aircraft and flying bombs.

By this time, the United States had officially entered the War, and with MIT's Differential Analyzer to solve the ballistic shell differential equations and the acronymic mainframe computers to calculate the firing tables, together with the new vacuum-tube proximity fuse that effectively detonated the anti-aircraft shell when close to the target based on the range calculations, the M-9 continuous feedback loop-controlled AA guns were ready for war.

The German V-1 flying bombs came by day and night. During the day, the only fighters that could challenge them were the fast low-flying Hawker Tempests, but close engagement ran the danger of self-destruction within the periphery of a successful V-1 bomb mid-air explosion and shock wave; stand-off machine gun bullets bounced off the thick plating of the missiles, and heavier anti-aircraft shells were difficult to target from range against the fast (550 km/h) V-1 rockets.

In the ultimate of hell-bent daring, RAF pilots flew over the English Channel, and from behind the approaching V-1s, diving to increase speed, they carefully positioned their wingtip to within 15 cm below the V-1 airfoil, causing the bottom-side air pressure to suddenly increase and the flying bomb to pitch, yaw, and roll in accord with the aerodynamic Bernoulli effect (unequal air pressure on the top and bottom of wings that produces lift). The sudden orientation change would override the V-1's pitch- and yaw-control gyroscopes and the rocket would dive and detonate at sea; it was estimated that 16 V-1's were destroyed in this intrepid aerodynamic Bernoulli attack.[6]

However, there were 6,725 Nazi flying bombs coming by day and night in the June 1944 attacks, and daring pilot feats notwithstanding, expert systems saved the day in the Second Battle of Britain, for the M-9 Predictor and its progeny purportedly succeeded in targeting nine out of ten V-1 buzz bombs in a raid over Kent and London, demonstrating the prowess of the continuous feedback loop anti-aircraft gun-laying expert system.

Thus, radar and the *M-9 Predictor* were critical to overcoming the Nazi's *Sea Lion* cross-Channel invasion plans.

THE BIOLOGICAL NEURON NEXUS

Warren McCulloch was a neurophysiologist graduate from Yale Medical School with a specialty in epilepsy and head injuries. His work led to an appointment as head of the University of Illinois' psychiatric research laboratory, seemingly a long reach from anti-aircraft gun-laying.

But after attending a lecture by Wiener on adaptive feedback, he surprised himself with the thought that an electron does not constitute a current, but that a stream of electrons held together in a wire, or by electric or magnetic fields does produce a current of electrons with a "purpose" (for example turning on a lamp).

Therefore, in analogy, although a single neuron in the brain when stimulated has no *sense* in and of itself, an aggregate of neurons and their synaptic connections can form a sense pattern that acts as a *feedback* with a purpose; that is, a "thought".

The idea then is to mimic the brain with *artificial neurons* that when activated in an *artificial neural network* will constitute a sense that the stimulus and the feedback response are recognized and stored, just like a thought in the human brain.

With this idea in mind, McCulloch sought the help of the 18-year-old mathematics prodigy Walter Pitts, who noted that since the neurons' activation was either *on* or *off*, they could be binary coded using Boolean algebra, and Claude Shannon's two-state on/off electronic switches could form logic gate cascades to process the *sense*, and store sense patterns, thereby *learning* through cybernetics.

In order to improve accuracy and hasten learning, McCulloch's basic artificial neural network (ANN) needed more sophisticated artificial neuron activations. In the late 1950s in another example of Wiener's interdisciplinary nexus, Cornell psychopathologist Frank Rosenblatt made the then solely on/off artificial neuron more impressionable by attaching a *weight* to each artificial neuron activation to modulate the synaptic patterns, and he called his parameterizing artificial neuron activations idea a *perceptron*.

Rosenblatt's perceptron was first tried in *computer vision*. A bank of photoelectric cells focused on a test image of two squares. The reflected light from the image was converted into analog electrical signals by the photoelectric effect, and those signals were digitized for greyscale-mapping onto a pixel matrix stored in the memory of an IBM 704 computer.

By weighting each pixel's activation to reflect the intensities of the light and dark patterns of the test pattern, *parameterization* iteration could produce an accurate rendition of the target squares in the computer's memory, producing the first instance of artificial neural network pattern recognition.

It is important to realize that the IBM 704 has not just reproduced the image to display on a video screen like a TV camera, its artificial neural network has *recognized* the features of two squares for storage in memory, and once having done so, the perceptron network has *learned* to identify and classify two-square patterns.

However, in his 1969 book *Perceptrons*, AI pioneer Marvin Minsky argued that the perceptrons' combinatory binary circuits could not perform the logical XOR, and would thus be limited in application. Minsky's belief was later disproved by a layered network of perceptrons, but not before it chilled AI neural network research into a ten-year "first AI Winter" that thawed only to encounter the ill-fated 1981 Japanese Fifth Generation comprehensive expert system project to resource and export *knowledge*, which failure after ten years engendered the "second AI Winter" into the early 1990s.[7]

Nonetheless, the *perceptron* persevered as an artificial neuron network that takes multiple binary (0 or 1) inputs x_i that are analogous to on/off activations of neurons and produces a single binary output decision a (*Yes* = 1 or *No* = 0). Each x_i input is multiplied by a weight w_i reflecting the significance of that input to the task at hand, the weighted inputs are summed, and if the weighted sum is less than or greater than or equal to some chosen threshold bias b, the decision a to activate or not is made, as illustrated by Figure 9.1.[8]

For example, you are trying to decide whether to attend a piano recital; there are a few decisional factors but it is a binary decision, *yes* or *no*. Say the input factors are:

x_1 = close by so you can walk

x_2 = girlfriend goes with you

x_3 = girlfriend's little brother also comes

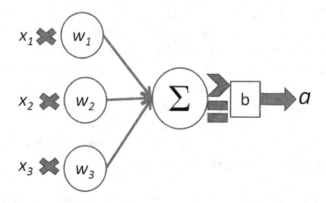

FIGURE 9.1 The perceptron.

You analyze the situation: You like piano music, the recital is close by, and you hope that your girlfriend would go with you, *but* she may bring her little brother along. In that case do you still want to go? Is any one factor decisive? If not, then a deeper analysis is required. The x_i input factors (activations) are known and the relative importance of each input factor can be determined by attaching a weight to it; for instance, having no car and little money, within walking distance is important, so $w_1 = 3$, your girlfriend coming is very important, so $w_2 = 5$, but her little brother also coming is a definite negative, so $w_3 = -4$. You like piano music so the threshold for attending is a low (for biases) $b = 6$.

Therefore, if you can walk to the concert and your girlfriend will come without her little brother, the weighted sum is $1 \times 3 + 1 \times 5 + 0 \times (-4) = 8 > 6$, where the little brother factor x_3, has been set to zero, and so the decision will be to go, you can walk with only her to the recital. If she and her little brother will be out of town and not able to come with you, then you won't go even if you can walk to the recital since the inputs x_2 and x_3 have been extinguished, so $1 \times 3 + 0 \times 5 + 0 \times (-4) = 3$ is below the threshold, so you won't go even when the concert hall is close by.

If your girlfriend will go but brings her little brother, your decision is more difficult, but the perceptron will decide for you. The weighted sum with an activated x_3 is $1 \times 3 + 1 \times 5 + 1 \times (-4) = 4 < 6$ reflecting the importance of x_3, so you will not go if her little brother tags along.

The weighting clearly helps decision-making, but the perceptron has also identified a critical factor, namely the little brother. Your decision can be helped along by finding some ground truth, for instance, if little brother has violin lessons that night. The worst case is of course your girlfriend does not come with you, but her little brother does, you have to pay for an Uber ride, and the pieces played at the recital are all atonal. This reflects the importance of investigating the *ground truth*.

Interestingly, in addition to deciding for you whether to go or not, the perceptron also reveals an *inference* that piano music may not be all that important to you; that is, your girlfriend's unfettered accompaniment is likely more important than any cultural pretensions you may have.

If you find out that the recital will have pieces by composers that you particularly like, you can lower the threshold towards attending; that is, the bias towards attending is increased, perhaps even to the extent of little brother tagging along being worth it.

The perceptron therefore is a sophisticated decision-making device operated by assigning weights that signify the relative positive and negative significance of decisional factors, with a threshold bias reflecting the importance of the decision.

Assigning weights is called the *parameterization* of artificial neural networks, which is an iteration achieved by gradient descent and back-propagation, to be discussed in the next chapter.

NOTES

1. Birkhoff, although an admitted right-wing conservative, averred that he was only promoting the advancement of home-grown American mathematicians and physicists in rejecting a position for the European Einstein, but Wiener was an American and both he and Einstein were Jewish.
2. Bombers flew in formation, and even though fighter planes would often stay in wing formations for both attack and defense, once broken off the wing, dog-fight maneuvers were not predictable
3. N. Wiener 1948, *Cybenetics*, Technology Press; N. Wiener 1950 (1988), *The Human Use of Human Beings*, Da Capo Press.
4. Quote of Violet Regan, the wife of a member of the Heavy Rescue Squad in Millwall, J. Gardner 2011, *The Blitz*, Harper.
5. Women were thought to be more thorough and less prone to error than men, and according to AA Command, spinsters were the best of them all at the computations.
6. Thomas, A., 2013, *V1 Flying Bomb Aces*, Osprey Publishing. The aerodynamic Bernoulli effect is the basis of aviation, the rounded at the top and flat-bottom shape of an airfoil causes the air stream at the angle of attack to travel a longer distance over the top, decreasing the air flow density and therefore the pressure compared to the bottom of the airfoil, thereby achieving lift.
7. Ref. Minsky, M., & S. Papert 1969, *Perceptrons, an Introduction to Complex Geometry*, MIT Press. An XOR binary gate outputs *True* when only one of the inputs is *True*, one or the other but not both.
8. Ref. Rosenblatt, F. 1958, *Psychological Review*, Vol. 65, 6. Example adapted from Nielsen, M., 2018 , *Neural Networks and Deep Learning*, academia.edu online pdf. Perceptrons can be implemented as NAND gates, so they can be programmed in integrated circuit (IC) design.

CHAPTER 10

Parameterization

I F PERCEPTRONS ARE LAYERED and connected layer-by-layer, an artificial neural network (ANN) is formed. In Figure 10.1, a simple 4×4
matrix is shown, but inapposite to the typical matrix nomenclature, in
feed-forward mode every layer is horizontally proceeding left to right and
counted, and every artificial neuron's activation a_j is vertically proceeding
downward and counted ($j = 0,1,2,3$) and is called a *row* and not a column
(as in matrices).

Every row has four nodes which are connected to the nodes of the preceding and succeeding layers; the first row of neurons constitutes an *input
layer* that receives stimuli or information from external sources; the next
two layers are the *hidden layers*, and the last row is the *output layer*, which
will be flattened out to form a *decision vector* by simply stacking the rows
on top of each other. The indices of the nodes $a_j(L)$ are shown in Figure.
10.2. The ANNs are volume matrices where the third dimension typically
carries the colors.

In the perceptron example of the previous chapter, the weights and
biases were assigned according to the individual's preferences, but just
how are the weights and biases determined for any given situation by a
machine that initially and ostensibly has no preferences?

In artificial neural networks, the initial activations (*initialization*)
are not determined from some preferences, but rather from a random
(for example, Gaussian) distribution, and this is the "blank" canvas of
artificial neuron activations that are modulated by a weighting process
called *parameterization*, that in machine learning fits data from training
datasets.[1]

DOI: 10.1201/9781003463542-10

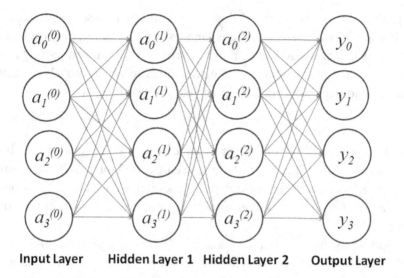

Input Layer   Hidden Layer 1 Hidden Layer 2   Output Layer

FIGURE 10.1   4 × 4 artificial neural network.

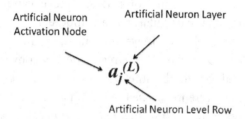

FIGURE 10.2   Artificial neuron indices.

Parameterization provides greater significance to distinctive features by enhancing artificial neuron activations, or conversely diminishing the indistinguishing features, by multiplying the activation $a_j(L)$ by a weighting factor and adding a bias to the sum of activations in the layer of neurons. The weighting factors and biases may also be hand-engineered suitable for a desired machine learning effect.

An artificial neural network is thus a multilayer perceptron matrix array. In a *fully connected* ANN, the weighted activation of all the neurons in layer (0) of the network is connected to every neuron in the succeeding layer (1) and so on; for example, the activation of the 0th neuron in the succeeding layer (1) is given by the sum of the ith row and jth layer weights $w_{i,j}$ times the activations of the preceding 0th $i$layer's j neurons,

$$a_0^{(1)} = w_{0,0}a_0^{(0)} + w_{0,1}a_1^{(0)} + w_{0,2}a_2^{(0)} + \ldots$$

To enhance features in the training set data, different biases (b_0) are added to the sum of weighted activations for a level, for instance, positive biases to ensure that the weighted sum in a neuron layer will meaningfully contribute to the feature extraction of distinct features, or conversely negative biases to downplay or totally ignore irrelevant features and noise,

$$a_0^{(1)} = w_{0,0}a_0^{(0)} + w_{0,1}a_1^{(0)} + w_{0,2}a_2^{(0)} + \ldots + b^{(0)}.$$

The binary "on" (1) or "off" (0) of biological cortical neurons and the linear increase of say light intensity in the semi-closed interval $[0, \infty)$ would require ANN to have an unlimited range of intensities, but ANN is only interested in *differences* of intensities to capture the essence of the training set data.

Optical displays like liquid crystal televisions use a *greyscale* of *decimal* values between 0 and 1 in the closed interval $[0, 1]$ that can have *unlimited differences* of weighted activation intensities. That is, because in the number theory *number lines* of decimal fractions within a range are infinitesimally small, the weighted artificial neuron activation level can be extremely fine yet still distinguishable (particularly by a computer).

Furthermore, small changes in weights and biases in a closed interval $[0, 1]$ will iteratively produce smaller changes as ANN converges to the input data, and there will be no absolute large changes in weighted sums and biases that may inadvertently cause the ANN parameterization to diverge.

Greyscale is generated by the *sigmoid function* (also called the *logistic function*) that restricts variables x to the closed domain $[0, 1]$ to produce a countably infinite number of possible values between 1 and 0,

$$Sigmoid\ Function = \sigma(x) = \frac{1}{1+e^{-x}}$$

To express the probabilities used in pattern recognition, the sigmoid function output must be positive because there is no such thing as a negative probability; the exponential function e has no negative values, so it is ideally suited to provide positive greyscale values.[2]

The sigmoid function σ can also represent a smoothed version of a step function, so it can also classify, depending on the value of x, as either negative or positive, as can be seen by the plot of $\sigma(x)$ in Figure. 10.3, and thus objects can be separated into the left and right sides of the graph, and so used for object classification, such as distinguishing cats from dogs and personal from spam emails and classifying hip-hop and classical music aficionados in different taste clusters.

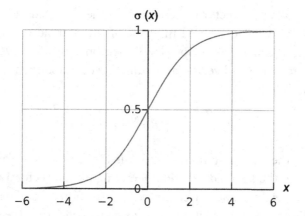

FIGURE 10.3   The sigmoid function.

The sigmoid function for one neuron in one layer operates on the neuron activation weights and biases as

$$\sigma\left(w_{0,0}a_0^{(0)} + w_{0,1}a_1^{(0)} + w_{0,2}a_2^{(0)} + \ldots + b^{(0)}\right)$$

$$= \frac{1}{1 - e^{-\left(w_{0,0}a_0^{(0)} + w_{0,1}a_1^{(0)} + w_{0,2}a_2^{(0)} + \ldots + b^{(0)}\right)}}$$

There are multiple neurons in each layer of a neural network, so a given neuron receiving a direct or synaptic greyscale activation from a preceding layer's neurons itself has an activation level that can be represented by a weighting matrix multiplication of the preceding layer neuron activation level vector, plus the bias vector, and then the whole kit operated on by the sigmoid function.

For example, the greyscale activation levels for the neurons in layer (1) as a function of the neuron activation levels of layer (0) in the simple 4 × 4 neural network of Figure 10.1 in matrix form is given by

$$a^{(1)} = \sigma \left\{ \begin{bmatrix} w_{0,0} & w_{01} & w_{0,2} & w_{0,3} \\ w_{1,0} & w_{1,1} & w_{1,2} & w_{1,3} \\ w_{2,0} & w_{2,1} & w_{2,2} & w_{2,3} \\ w_{3,0} & w_{3,1} & w_{3,2} & w_{3,3} \end{bmatrix} \begin{bmatrix} a_0^{(0)} \\ a_1^{(0)} \\ a_2^{(0)} \\ a_3^{(0)} \end{bmatrix} + \begin{bmatrix} b_0^{(0)} \\ b_1^{(0)} \\ b_2^{(0)} \\ b_3^{(0)} \end{bmatrix} \right\}$$

where $a^{(1)}$ is a layer 1 vector representing all the weighted neuron activation levels in layer 1 expressed in terms of the sigmoid function of the weights matrix times the previous layer's neuron activation levels plus the biases for each weighted vector layer sum. The equation then can be written compactly as[3]

$$a^{(1)} = \sigma\left(\boldsymbol{W}a^{(0)} + b^{(0)}\right) \tag{10.1}$$

where $\boldsymbol{W}$ is the *weighting matrix* and b is a column vector of biases in layer (0), with the sigmoid function operating on the vector $\boldsymbol{W}a^{(0)} + b^{(0)}$ to produce the weighted and biased greyscale activation $a^{(1)}$ of the neurons in layer 1. This equation explicitly shows the dependence of succeeding layer neuron activations on the weighted neuron activation levels and their biases of all the neurons in the preceding layer.

Seen in this way, each neuron layer is a perceptron arrayed in a multilayered network called appropriately enough a multilayer perceptron (MLP) model, the earliest modern artificial neural network. Each neuron has an input comprised of the weighted and biased neuron activation levels from the preceding layer, and a resultant activation of its own, which will be subsequently modulated by weights and biases for input to the neurons in the next succeeding layer in the so-called *feed-forward* mode.

The activation of layer (1) neurons in terms of the sigmoid function is

$$a^{(1)} = \sigma\left(\boldsymbol{W}a^{(0)} + b^{(0)}\right) = \frac{1}{1 - e^{-\left(\boldsymbol{W}a^{(0)} + b^{(0)}\right)}}$$

The activation of every layer's neurons is calculated as described above, and since artificial neural networks may have many layers and neurons, together with voluminous training set data, even though computationally burdensome, such a network is eminently programmable because massively parallel GPU computer processing is very good at the linear algebra of matrices.

Such originally daunting computations show why artificial intelligence could not really take off until the advent of mass-storage, ultra-fast parallel-processing computers, and the creation of efficient learning and recognition algorithms taking advantage of that hardware.

The sigmoid function and the hyperbolic tangent function (tanh) were and are commonly used in AI systems, but the later-employed and simpler rectified linear unit (ReLU),

$$f(x) = max\,(0, x),$$

which is just a straight 45° line that changes all the negative activations to 0, and with a *softmax* function later in the network to provide greyscale. This simple function is obviously easier to compute than sigmoid and hyperbolic tangent and therefore faster while making no demonstrably significant differences in output accuracy compared to other greyscale functions for many ANN algorithms.

Furthermore, since the learning gradient approaches zero when the sigmoid and tanh neurons activation level outputs are near either 0 or 1, they bunch up smaller and smaller at the extremes of the interval [0,1], and learning may severely slow down or stop (saturate).

An open-ended ReLU neural network parameterization will never saturate, so the learning will not slow down; however, if the final weighted input to a ReLU neuron network is negative, the gradient vanishes (goes to zero), and the ReLU neuron network will stop learning altogether. These imperfections have spawned a plethora of alternate greyscale conversion functions and *hyperparameters* which can be employed as needed for specific algorithmic tasks.[4]

Among them is the *softmax* function, true to its name, is a *max* function that *softens* say a *ReLU* by taming monotonically increasing numbers into the closed interval [0, 1] to provide greyscale,

$$Softmax = \xi\left(z\right)_i = \frac{e^{z_i}}{\sum_{j=1}^{K} e^{z_j}}$$

where z_i is the ith element of the vector z, and $\xi(z)_i$ is normalized by dividing each exponential function by the sum of the exponential functions to ensure that the values are between 0 and 1 as required for probabilities; $\xi(z)_i$ cannot be negative because the exponential function cannot be negative.

Choosing among sigmoid, tanh, ReLU, and others, and *hyperparameters* (Chapter 17) is an exercise in first addressing the task at hand in its simplest implementation, and then as problems arise, finding the best corrective functions and procedures, usually simply by trying out and comparing results. Where improved classification and prediction accuracy are required, algorithm tuning, different cost functions to hasten convergence, hyperparameterization, and so on can be called upon, all the while keeping in mind increased computational burden.[5]

Procedural choices often are made by just going with what works best and not necessarily completely understanding why something works better than something else.

The parameterization of the artificial neuron activation provides a means for the dynamic adjustment of synaptic patterns for feature extraction, but how exactly are the weights and biases calculated so that the artificial neural network can learn and recognize, thus exhibiting artificial intelligence?

NOTES

1. The types of initial distributions should be varied to avoid the same initialization predispositions of the ANN from the that may affect the learning by parameterization.
2. Like the sigmoid function, *fuzzy logic* expresses degrees of truthfulness by assigning values between 0 and 1 instead of the binary 0 or 1.
3. The "columns" in ANNs are the layers going left to right, which is often confusing because layers are usually represented by vertical stacks.
4. ReLU alleviates the sigmoid function vanishing gradient problem because the deeper layers train very slowly due to the exponential gradient decrease to values so small that they do not change the weights. Other greyscale conversion functions are CUBE, ELU, HARDSIGMOID, HARDTANH, IDENTITY, LEAKY RELU, RATIONAL TANH, RRELU, SOFTMAX, SOFTPLUS, and SOFTSIGN.
5. For cost functions, see Chapter 11; for hyperparameterization, see Chapter 17.

CHAPTER **11**

Gradient Descent and Backpropagation

AN ARTIFICIAL NEURAL NETWORK (ANN) must be trained to recognize and classify objects and data; young ANN is learning how to recognize by being presented with training sets of images and data. The neuron activation levels have been parameterized initially by a Gaussian distribution of weights and biases, her neural network is thus a blank canvas, and she is ready to *machine learn.*[1]

The difference between her initial random artificial neural activation distribution and the training set data usually will be quite large, and her learning task then is to reduce that difference in order to learn the features of the training set images and data, store them in memory, and then recognize the images and data presented to her.

That difference is measured by a *Loss Function*, expressed by the sum of the squares of the differences between the outputted activations of the decisional row vector neurons a in layer L and the corresponding training set y vector,[2]

$$Loss\ Function = (a^{(L)} - y)^2.$$

As employed in artificial intelligence, taking the average of all of the Loss Functions over all the m samples in the training set gives the *Average Cost Function*; it comprises all the weights and all the biases in the network averaged over all the samples i in the training set.

DOI: 10.1201/9781003463542-11

The Average Cost Function $\overline{C}$ is a term taken from economics as the Cost of inefficient production; C can also be thought of as just the *Cost of Being Wrong* in regard to matching the ANN activation pattern to the training dataset, so the minimization of the Average Cost Function will reduce the differences between ANN's neuron activation patterns and the training dataset samples.

Calculating the Average Cost Function will not be as computationally burdensome as calculating the Loss Function of each training set sample in turn, and it is noted that taking the average will beneficially average out noise and irrelevant factors that may be in the data.

The Average Cost Function then is defined as[3]

$$Average\ Cost\ Function = \overline{C} = \frac{1}{2m} \sum_{i=1}^{m} \left[(a^{(L)} - y)^2 \right]_i,$$

where recall from Chapter 10, Eq. 10.1, that

$$a^{(L)} = \sigma \left(W a^{(L-1)} + b^{(L-1)} \right)$$

In elementary calculus, minimization of any continuous and differentiable function of a single variable is easily done by taking the derivative with respect to that variable and setting it equal to zero to find where the slope is horizontal; that is, zero slope, which is the minimum point of a concave curve.

Taking C to represent any Cost Function, a quadratic C is continuous and differentiable, so it can be differentiated, for example, with respect to a weighting parameter variable w,

$$Set\ \frac{\partial C}{\partial w} = 0$$

To find the minimum point with respect to a weighting variable w, choose a point on the curve and determine what direction to move in order to reach the minimum point on the C–w curve. This direction is given by the sign of the slope of the curve at that chosen point; the analogy is a ball rolling on the Average Cost Function curve; it will roll left if the slope is positive and right if the slope is negative and with a speed depending on the severity of the slope. If not rolling too fast, it will stop at the horizontal slope minimum, thereby determining a minimum *Cost*.

This gives the clue that this operation would be best described using a vector since it has both magnitude and direction. From a randomly chosen point on the Average Cost Function versus weighting variable (*C* vs. *w*) curve, as the minimum point is approached very closely, since the absolute value of the slope decreases and the rolling ball slows down, the learning algorithm can accordingly reduce the step size of the ball to prevent inadvertently overshooting that minimum point.

Unfortunately, the Cost curve with respect to the variable *w* may have many concavities and therefore multiple *local minima* and also include convex curve maxima (that also have zero slope) as shown in Figure 11.1, left, but the Cost will not be minimized with respect to the weighting variable unless the *global minimum*, meaning the absolute minimum of the Average Cost Function, and not just a *local minima*, is found. The problem of encountering local minima in the search for the global minimum has plagued ANN artificial intelligence research since its inception.

If the minimization of the Cost Function for two variable weights is to be determined, instead of *C* versus *w* curves there will be a three-dimensional contour of Cost Function versus two weights as shown in Figure 11.1, right, with a rolling ball searching for the minimum of the contour.

Local and global minima for two-dimensional and three-dimensional models can be found by simply looking at the computer-generated *C–w* plots, but artificial neural networks usually have many, many weight and bias parameters (ChatGPT has trillions), so the Cost will be a function of

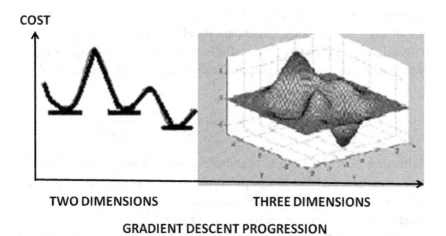

COST

TWO DIMENSIONS   THREE DIMENSIONS

GRADIENT DESCENT PROGRESSION

FIGURE 11.1   Cost Function with respect to weighting curves in 2D and 3D.

many, many variables, and any contour of greater than three dimensions is of course impossible to visualize.

Although humans cannot physically visualize in greater than three dimensions, they can conceptually view with the mind's eye and, fortunately, the computer can operate in almost infinite dimensions; so the computation of Cost Function minimization with respect to *any* number of weighting variables can be performed by finding the *gradient* of vector analysis.

The gradient is a vector representing the direction of the steepest *descent*, with a magnitude that is a measure of the steepness of the multidimensional contour, making it ideal for finding the Cost minimum (since it will be finding a minimum, it will be a gradient *descent* and thus a *negative* gradient vector),

$$\text{Vector of Steepest Descent} = -\textbf{grad}C = -\nabla C$$

The gradient operating on the Cost Function C in an example for just the weighting variables in three dimensions is

$$-\nabla C = -\left(\frac{\partial C}{\partial w_x} \textbf{\textit{i}} + \frac{\partial C}{\partial w_y} \textbf{\textit{j}} + \frac{\partial C}{\partial w_z} \textbf{\textit{k}} \right)$$

where $\textbf{\textit{i}}$, $\textbf{\textit{j}}$, and $\textbf{\textit{k}}$ are the unit vectors in the x, y, and z directions, respectively, and w_i are the weights in those contour directions, respectively.

A ball rolling on the Cost Function contour will roll toward minima, and although it is impossible to visualize, a greater than three multidimensional contour still can be described mathematically with the same ball analogy, so a gradient dependent on all the weight parameters in an artificial neural network of many dimensions j in "directions" designated by the *unit directional* vectors $\textbf{\textit{e}}_j$ in principle can always be calculated as

$$-\nabla C = -\left(\frac{\partial C}{\partial w_1} \textbf{\textit{e}}_1 + \frac{\partial C}{\partial w_2} \textbf{\textit{e}}_2 + \frac{\partial C}{\partial w_3} \textbf{\textit{e}}_3 + \ldots + \frac{\partial C}{\partial w_j} \textbf{\textit{e}}_j \right)$$

Multivariate calculus is a good example of how mathematics can broaden one's mind beyond what can be perceived by the senses. That is, "seeing is believing" is helpful, but neither necessary nor complete in mathematics where "seeing" means conceptual contemplation.

This is one of the reasons why *doing mathematics* may indeed be the touchstone of intelligence, for no one can doubt that the creativity and construct of the complex conjugate vector calculus of a *complete inner product Hilbert space* providing a scalar distance function requires no little gray matter.[4]

The algorithm to compute the minimization of the Cost Function in multivariate calculus computes the gradient vector at a point on the multidimensional contour by taking a step in the gradient direction and repeating the calculation of the gradient vector over and over again until a minimum is found. The size of the step is called the hyperparameter *learning rate*; a large step for fast learning will hasten the process to convergence and reduce computational burden, but too large a step may overstep a minimum.

The gradient descent of the Cost Function will involve taking the derivatives of the sigmoid (or other greyscale conversion) function σ, the weighting matrix W, the previous layer activation levels $a^{(L-1)}$, and the previous layer biases $b^{(L-1)}$, altogether incrementing the values of the weights and biases that will cause the most rapid minimization of the Cost Function.

The sign of the resulting adjusted weights and biases promoting Cost minimization indicates a higher (+) or lower (−) adjustment, and the relative magnitudes of the weights and biases reveal which of the adjustments will have the greatest impact in reducing the Cost, in other words, how sensitive the Cost is to a particular parameter adjustment.

The gradient descent therefore amazingly encodes the relative significance of each weight and bias toward minimizing the error with a labeled training dataset, in effect teaching the artificial neural network how to accurately recognize the training set data for what it represents.

Since the derivative of the gradient shows how the gradient is changing, as its absolute value decreases, it must be nearing a minimum, so the second derivative of the Cost can be arrayed in a *Hessian matrix*, from which the determinant of a second-order partial differential equation with respect to the scalar field of weights and biases can be calculated. Because the determinant is positive definite, it can be used to test for extrema; a minimum if the Hessian determinant is positive, if negative, it is a maximum, and if zero, a saddle point of a hyperbolic paraboloid.

The fearsome process of calculating second derivatives thankfully has some labor-saving techniques such as the gradient descent acting to change the *velocity* (already a derivative) instead of *position*, which avoids the large second-order derivative calculations by taking a *momentum* with *friction*, reducing the velocity as minima are approached in small steps from different points on the curve in a physics-oriented approach.[5]

Gradient descent and/or the Hessian will find the first minimum encountered but will stop when the slope at that minimum is zero, essentially stopping the learning process, and unfortunately if it is a local minimum, foundering in a local minimum hollow. At this point, it will be necessary to start the calculation from that point once again to find the next minimum, hopefully, a *global* minimum instead of just another provincial local minimum.

BACKPROPAGATION

The process of gradient descent to minimize the Cost Function employs the simple minimization calculus. But the artificial neural network (ANN) forms synaptic connection patterns at each layer and passes them on to the succeeding layer. Therefore, gradient descent must operate throughout the layers of the artificial neural network *with reference to the preceding layers*. This is done by performing gradient descent going backwards through the network, in a process called *backpropagation*, essentially a feedback loop adjusting the weights and biases to minimize the Cost Function through the calculation of the gradient descent at each layer in terms of the preceding layer.

The significance of the activation level of each artificial neuron towards matching the training dataset can be increased or decreased by changing their weights and biases, and since the weighted activation levels of the preceding layer neurons will affect the weighted activation levels of a given layer's neurons, going backwards through the neural network layer-by-layer and recursively adjusting the weight and bias parameters in each preceding layer in accord with minimizing the Cost of Being Wrong, in principle, will ultimately match the artificial neural network synaptic pattern to the training set data pattern.

If the algorithm requires considerable computational power, the training data may be divided into *mini-batches* and run in turn as iterative *stochastic gradient descents* with validation and test sets to enhance the network's accuracy in matching the training set data.

Gradient descent backpropagation is based on the Gauss–Newton numerical analysis computation of non-linear partial differential equations called *Newton's Method* where "Newton" refers to numerically taking the derivatives conceived by Isaac Newton, to calculate the trend of the differentiation. The process of backpropagation is based on the fundamental *chain rule* of differential calculus, as will be seen below.

The *Loss Function* for a given artificial neural network layer L is

$$Loss\ Function = (a^{(L)} - y)^2,$$

where $a^{(L)}$ is the activation vector of the row neurons in layer L and y is the training set vector.

Since this will be the *Cost of Being Wrong* for each layer L, to avoid confusion with the L designating the network layer, and further to be in accord with the AI literature, C will be used for the Cost Function per layer. It is understood that the Average Cost Function over all the training set examples is designated by $\bar{C}$.

As is often done in mathematical derivations, a change of variable makes life easier, so define a new variable $z^{(L)}$ in terms of the weights $w^{(L)}$ and the biases $b^{(L)}$ in layer L, and the activation of the neurons $a^{(L-1)}$ in the previous layer $(L - 1)$,

$$z^{(L)} = w^{(L)} a^{(L-1)} + b^{(L)}$$

Recall that the neuron activation of a neuron, its weight, and bias in layer L is converted to greyscale by operation of the sigmoid function,

$$a^{(L)} = \sigma\left(z^{(L)}\right) = \frac{1}{1 + e^{-z^{(L)}}}$$

To first compute the sensitivity of Cost to the change in weighting factor variable, $\partial C / \partial w$, the fundamental chain rule of calculus is used, which is the mathematical basis of the idea of backpropagation,[6]

$$\frac{\partial C}{\partial w^{(L)}} = \frac{\partial z^{(L)}}{\partial w^{(L)}} \cdot \frac{\partial a^{(L)}}{\partial z^{(L)}} \cdot \frac{\partial C}{\partial a^{(L)}}$$

Taking the derivatives of each term starting with the last term (with a prime on the sigmoid function σ means taking the derivative, as shown in the second equation),

$$\frac{\partial C}{\partial a^{(L)}} = 2\left(a^{(L)} - y\right),$$

$$\frac{\partial a^{(L)}}{\partial z^{(L)}} = \sigma'\left(z^{(L)}\right) \equiv \frac{\partial}{\partial z^{(L)}}\left(\frac{1}{1 + e^{z^{(L)}}}\right),$$

$$\frac{\partial z^{(L)}}{\partial w^{(L)}} = a^{(L-1)}.$$

The last equation says that the change in $z^{(L)}$ with respect to the weight $w^{(L)}$ in layer L depends on the activation intensity of the neuron in the preceding layer, $a^{(L-1)}$; that is, as in the synaptic patterns of biological brains, the neurons that fire together are wired together, and in the backpropagation of artificial neural networks, the artificial neurons that fire together are chained together. So,

$$\frac{\partial C}{\partial w^{(L)}} = a^{(L-1)} \cdot \sigma'\left(z^{(L)}\right) \cdot 2\left(a^{(L)} - y\right).$$

Now average the Costs with respect to the weights in each layer L over the m training set samples,

$$\frac{\overline{\partial C}}{\partial w^{(L)}} = \frac{1}{m}\sum_{k=0}^{m-1}\frac{\partial C_k}{\partial w^{(L)}}$$

This gives the average of a dataset for level L of the derivative of the Average Cost Function with respect to the weight for that level.

Repeating the process for the rate of change of Cost with respect to the biases $\partial C / \partial b$ gives,

$$\frac{\partial C}{\partial b^{(L)}} = \frac{\partial z^{(L)}}{\partial b^{(L)}} \cdot \frac{\partial a^{(L)}}{\partial z^{(L)}} \cdot \frac{\partial C}{\partial a^{(L)}}$$

But from the definition of the new variable $z^{(L)}$ given before,

$$\frac{\partial z^{(L)}}{\partial b^{(L)}} = 1$$

and since the other terms have been determined above,

$$\frac{\partial C}{\partial b^{(L)}} = \sigma'\left(z^{(L)}\right) \cdot 2\left(a^{(L)} - y\right),$$

and recall that

$$\sigma'\left(z^{(L)}\right) = \frac{\partial}{\partial z^{(L)}}\left(\frac{1}{1+e^{z^{(L)}}}\right)$$

Now just repeat the process iterating backwards through all the layers, one-by-one to minimize the Cost with respect to the weights and biases of the layers in turn.

To consider each and every neuron in the layers, just add row subscripts to the a's and two subscripts (for row j and column k) to the w's in the terms so that[7]

$$z_j^{(L)} = \ldots + w_{jk}^{(L)} a_k^{(L-1)} + \ldots$$

and

$$a_j^{(L)} = \sigma\left(z_j^{(L)}\right)$$

and the Cost over the j rows of neurons will be the sum over m training set examples,

$$C = \sum_{j=0}^{m_L-1} (a_j^{(L)} - y_j)^2$$

The chain rule expression is now as

$$\frac{\partial C}{\partial a_k^{(L-1)}} = \sum_{j=0}^{m_L-1} \frac{\partial z^{(L)}}{\partial a_k^{(L-1)}} \cdot \frac{\partial a_j^{(L)}}{\partial z^{(L)}} \cdot \frac{\partial C}{\partial a_j^{(L)}}$$

where the activation is now in terms of $L - 1$, meaning *backpropagating*. Sum the above expression over L for all the different layers. This is the same as for the single neuron example, except for the Cost with respect to the activations in Layer $L - 1$; that is, the $L - 1$ neurons influence the Cost through multiple different paths because of the multiple neurons in the layers, and they must be all added up.

Performing these derivatives for a given layer will adjust the weights and biases in relation to the preceding layer (backpropagating) in accord with minimizing the Cost, and the activation levels of the artificial neurons (hopefully) will converge to match the training set data.

Ostensibly a great deal of computation, all these calculations can be efficiently performed by freely accessible software computational programs and algorithms run on host-computer coding platforms such as GitHub and Red Hat.[8]

In summary, the chain rule gives expressions for the derivatives that determine each component of the gradient descent vector by repeatedly stepping downhill through the network layers on the steepest slope towards the minimization of the Cost Function by adjusting weights and biases in each layer in turn by going backwards through the network.

For understanding today's artificial intelligence, it is critical to realize that the artificial neural network *was not specifically told what features of the training dataset to learn or how to learn* those features; the network learned *all by itself* because it knows calculus; that is, the AI machine knows the minimization techniques and how to apply the chain rule of calculus for backpropagation.

This is the essence of bottom-up artificial intelligence, the AI machine was not programmed to perform specific tasks or recognize particular things from the top-down; rather within its hidden layers and through its algorithms, gradient descent, and backpropagation, the machine can learn and arrive at conclusions based on images and data in the training set. In other words, the AI machine learns and performs *autonomously*, and then through reinforcement and unsupervised learning (Chapter 16), it can improve its learning and recognition capabilities on its own.

NOTES

1. A normal (Gaussian) distribution takes random noise into consideration and the bias parameter considers systematic error.
2. The reason for taking the square of the difference in the Loss Function is that if the goal is to have the difference approach zero, and since a negative loss (profit?) is just as bad as a positive loss, taking the square of the difference considers both. Squaring also renders outliers more pronounced, eliminates the possibility of negative values, and is a measure of the statistical variance; furthermore, the derivative of a squared expression is also easy to calculate, leaving the difference intact as a factor. The absolute value of the difference is not used because it regresses to the median instead of the mean, and it is undefined at the origin, causing unneeded computational problems. Refer to any book on statistics for details, for example, Navidi, W., 2019, *Statistics for Scientists and Engineers*, McGraw-Hill Education.
3. The ½ factor in the Cost Function equation is for convenience as it cancels out during calculations because the derivative of a squared variable gives a factor of 2, rendering the form of the subsequent equations simpler.
4. Hilbert space is an infinite dimensional generalization of the Euclidean space of three dimensions that expands vector analysis to the measurement, for instance, of *distance* as a multidimensional vector inner (dot) product, and although that distance cannot be visualized, it is among many other invariant physical quantities in Nature, and very useful for mathematical physics calculations, particularly for the covariance required by Einstein's

relativity, see the author's book Chen, R.H., 2017, *Einstein's Relativity, the Special and General Theories with their Cosmology*, McGraw-Hill Education (Asia).

5. Ref. Nielsen, M., 2019, *Neural Networks and Deep Learning*, adademia.edu pdf.

6. The logic of the chain rule of calculus can be made clear by just cancelling each partial derivative numerator with the denominator of the following partial derivative.

7. In keeping with the notation formalism used in the AI literature, j denotes the row (output) and k the column (input) instead of the more natural i row (input) and j column (output) used earlier in describing the artificial neural network because the latter would require replacing the weight matrix with its transpose, thereby messing up the application of the weight matrix on the activation level. Ref. Nielsen, M., 2019, *Neural Networks and Deep Learning*, academia.edu pdf, p. 41 footnote.

8. For a manual demonstration of a backpropagation computation, refer to Mazur, M., "A step-by-step back propagation example", mattmazur.com.

CHAPTER 12

The Cross-Entropy Cost Function

A COMMON PROBLEM ENCOUNTERED IN the minimization of the Cost Function is that the gradient descent at times will suddenly slow down and even stop altogether while backpropagating through the hidden layers, never minimizing the Cost, and therefore ceasing to learn.

Learning by minimizing a quadratic Cost Function may be slow because when the neuron activation pattern is very different from the training set data, the Cost is very high, and the weights and bias parameter iterations will take more time when confronted with a massive difference with the training set data, or simply stop because the error has gone beyond the pale, and can never iteratively overcome the differences.

The minimization of the Cost Function is a subject of *information science* which studies the continuous transfer of information and how to increase the speed of transfer. The principal idea is the reduction of *information entropy*, which is defined as "the average amount of information conveyed by an event, when considering all possible outcomes"; or "the randomness of the data contained in a data set"; or "the average rate of the information from the transmission source to the receiver", and "the difference from the *expectation value* of the information arriving at the destination".

All the definitions will be used as aids to understanding the mysterious information entropy. If the information from the source transmitted to the receiver is very different from the expectation value, there will be an

 DOI: 10.1201/9781003463542-12

element of *surprise*, which is measured in units of *surprisal* because the information received is not what was *expected*. For example, if a goalie scores a goal, this event has high surprisal. High randomness means high information entropy, which in turn means the uncertainty of the transmission, and low randomness means low information entropy, which means accurate information transmission.

Curiously, a seemingly "bad" (high entropy) transmission provides more information, which is not necessarily "learning from mistakes" but surprisingly finding new information; and a seemingly "good" (low entropy) transmission does not provide any new information, so it is not necessarily good; so good or bad depends on the objective under study. For the artificial neural network, *learning* starts with bad high entropy and progresses to good low entropy.

In machine learning, the training set is the source and the information therein is processed by the artificial neural network (ANN) through gradient descent and backpropagation, and the Cost Function, being the difference between the information transferred and that received after processing by the ANN, decreases the information *entropy*.

A high information entropy may cause the Cost Function to slow down or stop; to overcome this problem requires a Cost Function that can accelerate gradient descent when encountering a large difference between the ANN and the training dataset. The *Cross-Entropy Cost Function* is based on reducing the "surprise" of the Cost Function upon reception of the information from the training set.

THERMODYNAMIC ENTROPY

To understand information entropy, one must first understand the mysterious *entropy* itself. The Second Law of Thermodynamics states that the thermal state of the Universe will always increase, entropy is a measure of disorder or uncertainty in a system, and for spontaneous changes, the disorder of the system will always *increase*; this is the *Law of Entropy Increase*.

This seemingly abstruse concept has many different aspects and deep mathematical expression, but it is often encountered in everyday life. For example, while walking, if your shoelaces loosen and start to disengage, they have been released from the ordered state constraint of being securely tied, and as you continue walking, your shoelaces will further unravel as your walking system continues to increase disorder. Your laces will never miraculously re-tie themselves as you walk, returning to a more ordered state, but rather the longer you walk, the looser your shoelaces become.

Within your walking system, the shoelaces have gone from the more-ordered tied state to the more *disordered* untied state, and the entropy of your walking system has increased.

Entropy in action can also be directly observed, for instance, in the array of computer cables and peripheral device wires below your desk; no matter how carefully initially arranged, at the next observation, they have all somehow mysteriously deteriorated into a hopeless tangle of high entropy disorder.

This spontaneous increase of entropy can be explained by the probability of occurrence of states in statistical mechanics. To illustrate, if two coins are tossed, the is a 1/2 probability of heads or tails; if five coins are tossed, the probability of coming up all heads or all tails is very low, and the probability of four heads and one tail (or vice-versa) is five times larger because five different arrangements of heads and one tail satisfy the criterion (since there are five different coins that can be the odd-out). For three heads (tails) and two tails (heads), there are ten cases each that satisfy the criterion (3-2 either way) and thus are ten times as likely to occur. So the highly-ordered state of all heads or all tails has only one state, and the more disordered state of three heads and two tails has ten possible states. Therefore, the high-disorder state is much more probable because there are more available states.

If 100 coins are tossed, the total number of different combinations of heads and tails is about 10^{30}, and thus the probability of coming up all heads is practically nil at $1/10^{30}$ because of the huge number of possible combinations.

And that is also why your untied shoelaces will not re-tie themselves as you continue walking and your computer cables are always in a tangled mess. The tied shoelaces and the carefully separated, parallel cables are more ordered states with fewer degrees of freedom, while the untied shoelaces and tangled cables have many possible different unordered states and are thus overwhelmingly more probable.

This also explains the frustration of never finding what you are looking for in places where you expect something to be; at the beginning of Summer, the possible places where your sunglasses might be are much greater than the one place where they happen to be.

Of course, re-tying your shoelaces and disentangling your computer cables make them more orderly, but the price to pay is increasing entropy in your body system (including the heat of annoyance) from expending energy to reorder those systems, and they will never exactly return to the same original state (*thermodynamic irreversibility*). The result is that even if you reorder the system, the total entropy of the coupled systems (you and the shoelaces and cables) has still increased.

The direction priority of the natural spontaneous transfer of heat (energy) is popularly expressed by the Second Law of Thermodynamics as "the direction of energy transformation is always from a hotter place towards a colder place". This seemingly obvious statement, however, can correct the misconception by many that leaving the refrigerator door open allows the cold air to come out when it really is the hot air going in and pushing the cold air out. A more vivid example is putting a kettle of cold water on the stove burner will not result in what little heat the water has being transferred to make the burner flame ever so little hotter; the truth of the Second Law is heralded by the kettle whistle.

With every spontaneous event, the thermodynamical arrow of time flies only forward, and as almost all events are irreversible, they always proceed towards the greater disorder; for example, if your freshly baked apple pie slid off the kitchen table and splattered on the floor in a disordered mess, you will not see it spontaneously reconstitute itself and fly back up to the tabletop in a more ordered state (except in film rewind humor). The event is also irreversible, as the apple pie cannot be reconstituted exactly as it was originally.

Entropy therefore is present in any system, natural or artificial and, in particular, communication. When a low-probability event is found in an element of a high-probability target vector, the error is large, and that event carries more disordered information than when a high-probability event is found because the latter event more closely matches the target vector. Correctly predicted events carry less information entropy because one already knows what is correct, and unexpected events are more disordered and carry more information entropy because it is informing one of something that was not expected (or appreciated).

Information theory was derived from electronic communications where clean signals are the expectation values. "Communication" is defined as "the identification of data from a source by means of the transmission of an encoded signal". Information entropy provides an absolute limit on the shortest possible average length of accurate expression of the data in the encoded signal, and if the entropy of the source is less than the transmission channel's entropy capacity, the information is deemed a "lossless" (good) communication.[1]

CROSS-ENTROPY COST FUNCTION

If the probability of some event E happening $p(E)$ is small, the information in the event is large (since it is improbable) and will cause a *surprise*, thereby providing unexpected information that can be good or bad; that is, in scientific research, an unexpected result may be a new discovery (good), or in machine learning, the ANN is not learning the training dataset (bad).

If the probability of some event happening is high, it was expected and therefore provides little or no new information because it was *expected*, so $p(E)$ is close to 1, and the (new) information transmitted in that event $I(E)$ is close to 0.

The founder of information science, Claude Shannon, uses the logarithmic function to describe this relationship because the logarithmic function is the only function where if the independent variable input is 1, the output is 0. The logarithmic function also increases very rapidly with the independent variable, in this case $p(E)$, and so is ideal for handling surprises and decreasing an extraordinarily high Cost.

If the probability of E occurring $p(E)$ is close to 0, the new information content $I(E)$ is close to 1 (previously unknown information), and the information entropy is very high; that is, a high surprise that may be revealing a new discovery, or that there may be a problem with the algorithmic system. The information and probability of the event relation can be written as

$$I(E) = -log_2[p(E)] = log_2[1/p(E)]$$

which clearly shows that the higher the probability $p(E)$ of the event E happening, the lower the information $I(E)$ transmitted in that event.

For artificial intelligence machine learning, however, a surprise is usually not welcome. It may cause the learning to slow down or altogether stop and may be because the artificial neural network activation patterns and the training dataset are so far apart as to cause "surprisal", and in the case where this surprise to ANN in the training dataset information transmitted is not meant to provide more information, but rather to be learned, as expected, and ANN is not learning it.

The Cross-Entropy Cost Function can employ Shannon's event information (training set data) as a function of probability according to the above equation.

Information science was derived from electronic communications, a clear signal which has the expected value at the receiver is an ideal *lossless* transmission. For machine learning, reaching the learning objective is lossless communication from the training dataset to the output layer of the ANN.

Although the mathematical theories behind information entropy are complicated, the idea of the Cross-Entropy Loss Function can be simply described. For an $M = 2$ binary classification (*Yes, No*), the Cross-Entropy Loss Function is

$$Cross\ Entropy \mid_{M=2} = -\left[y log(p) + (1-y) log(1-p) \right]$$

where y is the resultant binary indicator (0,1) and p is the predicted probability with the expression in square brackets being just the sum of the resultant times the predicted probability and the only alternative in this binary case, the probability of the only other possibility. The minus sign in front of the square brackets indicates the negative for a decreasing loss.

For multiclass classification $M > 2$ (for example, cats, dogs, hamsters, and horses), the Cross-Entropy Cost Function is the sum of the separate loss for each class label c per observation o,

$$Cross\ Entropy\ Cost = -\sum_{c=1}^{M} y_{o,c} log(p_{o,c})$$

where $y_{o,c}$ is the binary indicator (0,1) if the class label c is the correct classification for the observation o and $p_{o,c}$ is the predicted observation probability that o belongs to the class c.

It can be seen that the equation above can handle the surprises (such as the horse in the example above) in object classification by using the Shannon logarithm function, and then gradient descent and backpropagation to minimize the Cost, and have ANN converge to the classification task of the training dataset.

The Cross-Entropy Loss and Cost Function's information theory entropy underpinnings can be complex, but the implementation just involves writing a simple program representing the above equations or downloading canned software, such as *XENT Cross-Entropy*, letting it run, and seeing how well things progress, if at all.[2]

In many cases of artificial intelligence development, the "if it works, go ahead" engineering ethos trumps the "knowing why and how" of physics.

However, it is still true that if one wants to invent something new, or fundamentally improve something old, the theoretical bases of the technique will have to be understood.

But it is a fact that many of the recent successes of artificial neural networks have come about simply through experiment, trial-and-error, and heuristics, in a completely utilitarian manner that can be generalized, as expressed by the artificial intelligence pioneer Yann LeCun:[3]

> You have to realize that our theoretical tools are very weak. Sometimes, we have good mathematical intuitions for why a particular technique should work. Sometimes our intuition ends up being wrong. ... The questions become: how well does my method work on this particular problem, and how large is the set of problems on which it works well.

NOTES

1. Ref. Shannon, C., and W. Weaver 1971, *A Mathematical Theory of Communication*, University of Illinois Press.
2. Software packages for Cross-Entropy include XENT Cross-Entropy: Binary Classification, MCXENT: Multiclass Cross-Entropy, and RMSE_EXENT: RMSE Cross-Entropy.
3. LeCun quote from *Convolutional Nets and CIFAR-10: An Interview with Yann LeCun,* "No Free Hunch". December 22, 2014.

CHAPTER 13

Convolutional Neural Networks

IN HUMANS, VISION IS generated from light focused by the eye's lens onto a *retina* at the inside posterior of the eye; the retina converts the light into electrical signals that are received by an *optic nerve* behind the eye that transmits the signals through the *lateral geniculate nucleus* (LGN) relay pathway to an aggregation of neurons at the posterior region of the brain called the *visual cortex*. The visual cortex receives, processes, and integrates the vision signals to form a pattern for the brain to cognitively process by means of a network of neurons, one neuron of which is schematically illustrated in the middle of Figure 13.1.[1]

The *soma* of a cortical neuron is activated if the sum of the signals from other cortical neurons at the neuron's *dendrites* is greater than some threshold value. The neuron's activation is transmitted to other neurons through *axons*, whose *axon terminals* are connected to the dendrites of the other neurons, and in accord with the neuroscience maxim,

Neurons that Fire together, Wire together

Synaptic firing patterns of activated neurons are formed within the visual cortex which are then resolved by the brain to form images for cognition. Our thinking brains work by neural combinations, each neuron may be connected to up to 10,000 other neurons, passing signals to each other via as many as 1,000 trillion different synaptic path connection combinations.

DOI: 10.1201/9781003463542-13

In biological vision systems, initially un-activated cortical neurons will be activated when subject to a stimulus above some threshold energy; the neurons then may instigate synaptic connections to other neurons forming a sense of the stimulus.

In computer vision, a camera's lens receives photons of light reflected from an object and an array of sensors converts the light into electric signals by means of the photoelectric effect. The signals from this artificial retina are amplified and relayed in analogy with the optic nerve and LGN and transmitted to an *artificial neural network* (ANN) that is modeled after the visual cortex.

The ANN is a network of layers of arrays of artificial neurons. An example of a four-layer deep artificial neural network (with *layers* being the vertical columns) has an *input layer* to receive stimuli to the network, followed by two succeeding *hidden layers* and an *output layer*, each column layer having four row neurons all organized in a 4 × 4 array with node connections from each neuron in a given column layer to all the neurons in the succeeding column layer, as shown in Figure 10.1 of Chapter 10.

The artificial neural network's layers are held in computer memory as volume matrices (second-rank tensors) with the artificial neuron activation level as elements, typically with two dimensions for spatial distribution and one for color, and vectors (first-rank tensors) for decisional output, all within a neural network matrix (second-rank tensors).

The number, size, and type of hidden layers are determined by the recognition task. An artificial neural network with two or more hidden layers is considered to be a *deep neural network* (DNN). If all the artificial neurons in a succeeding layer are connected to each of the neurons in a preceding layer, the layers are said to be *fully connected*; if only some of the neurons in the preceding layer are connected to a succeeding layer, they form sub-matrix *windows* (called *filters* or *kernels*) in a *convolutional filter*

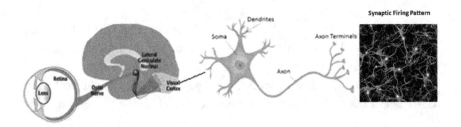

FIGURE 13.1   The human vision system.

layer whereby sectors of a preceding layer are selected for specific, finer or coarser, and positional *feature extraction*.

In *feedforward* mode, the activated neurons connect the artificial neurons in succeeding artificial neurons in the network to form synaptic patterns of activation, just as in a biological neural network.

In computer vision, a viewed image goes through a lens to a CMOS sensor, and from the photoelectric effect, photons are transformed into electrons to form an analog electrical current. An analog-to-digital converter (ADC) transforms the image to digital form as artificial neurons that are arrayed in an input layer in an artificial neural network computer matrix.

The human visual cortex has small clusters of neurons that are responsive to specific features or sectors in the visual field; these so-called *receptive fields* are then merged to constitute the entire image to the brain.

In a multilayer perceptron (MLP) neural network, all the neurons in adjacent layers are *fully connected*, and it therefore cannot employ smaller matrix filters to focus on specific areas of the visual field.

The hidden layers of a *convolutional neural network* (CNN) are not fully connected, but rather are interleaved with smaller windows of matrix *filters* akin to the receptive fields of the human visual cortex. In a typical CNN, a first *convolutional filter* sweeps over the input matrix layer and succeeding convolutional filters stride over the preceding hidden layer picking out features to constitute a *feature map* of the viewed object.

If there are hidden layers of smaller matrices than the input matrix to pick out features in the image, these are the *convolutional filters* that constitute the *inception layers* of a *convolutional neural network* (CNN), as shown in Figure 13.2.

A *convolution* is simply a mathematical operation of two functions to produce a third function that represents how the functions are conjoined. In two dimensions, a mapped function $f(x,y)$ is *convolved* by figuratively placing a *convolving* filter $h(x,y)$ over $f(x,y)$, spatially stepping (+1) through it, and integrating over area,

$$Convolution = \iint f(x,y)h(x+1,y+1)dxdy.$$

The double integral is over the filter's x times y area spanning a layer sector of the same size, the displacement shown is a +1 step (typically starting from the upper left-hand corner and first to the right, then down) of the $h(x + 1, y + 1)$ filter over the $f(x,y)$ as it strides step-by-step over successive

Computer Processed Convolutional Neural Network

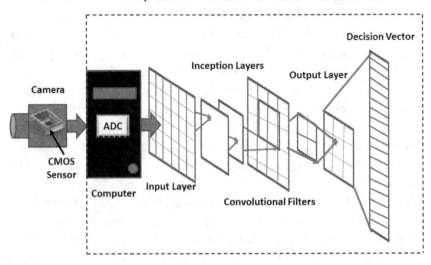

FIGURE 13.2   Computer vision artificial convolutional neural network.

same-sized sectors of the input matrix $f(x,y)$, extracting feature map matrices of the same size as the filter. The filter type, size, starting point, and step size are chosen to fit the computational task at hand.

The convolution process combines taking the *inner product* of the vector rows of two adjacent matrices and then integrating, which is a measure of the degree of *confluence* of adjacent vectors within the matrix.

The inner (or dot) product of two vectors is defined as the magnitudes of the vectors times the cosine of the angle θ between them. Also called the dot product, it can be illustrated for those who like basketball by the trajectory vector of the ball approaching the basket from the top of its arc with θ being the angle between the vector of the ball and the vertical axis vector through the hoop (Figure 13.3).

So if the ball is coming vertically down along the axis of the hoop ($\theta = 0°$ and $\cos \theta = 1$), it "sees" the full circular area of the hoop and thus has the maximum-sized target for a score. If the ball comes in horizontally ($\theta = 90°$ and $\cos \theta = 0$), it sees only the edge of the basket, and there is no possibility of a score. If the ball comes in at an angle of 60° (a relatively "flat" shot), $\cos \theta = 0.5$, it sees only half the area, so it has only half the chance of a vertically falling shot. Obviously, the smaller the angle of the ball's trajectory with the vertical, the more likely the ball will go through the hoop as measured by the cosine of the angle that determines the size of the target hoop as "seen" by the basketball.[2]

FIGURE 13.3   Basketball cosine.

Of course, shooting a ball almost straight up so it falls almost vertically through the net takes inordinate strength and, because of the long trajectory, is more difficult to control, so trading-off an angle of say 45° will give a 0.707 hoop area which is much better than a flat trajectory of say 75° which gives only a 0.26 hoop area target.

From this, it can be seen that the dot product is a measure of the magnitude of the *blending* of the two vectors to produce a scalar value. The inner product is a generalization of the dot product to multidimensional vector space (*Hilbert space*), obtained by multiplying the corresponding elements of the row vectors in a multidimensional matrix and summing the products to produce a scalar measure of the union of the multiple vectors constituting a new matrix.

Taking the double integral over the inner product of the matrices produces a *confluence* over area that measures the converging of the vectors in the matrices, just as in the merging of two flowing rivers, from whence the term came.

A convolutional filter striding over an artificial neural network matrix layer can be seen as a *sliding* inner product that extracts the confluences between the filter and the matrix layer, detecting, augmenting or dampening, and thereby extracting *features*.

As a filter matrix (also called a *kernel* or *window matrix*) glides over a matrix layer like a flashlight beam, it convolves the activation levels of the "illuminated" regions of the matrix, row by row. Prominent features will be enhanced because the elements of the *matrix layer* having higher-weighted activation levels will be affirmatively convolved by the inner product confluence and the higher-weighted activation level elements of the *filter*. Weaker features will be negatively convolved because of the inner product of small or negatively weighted activation levels in the matrix layer and the filter confluences will be small or even negative and thus can be overlooked as inconsequential or as noise.

The double integral produces a sum calculated by adding the inner products of the row vectors of the filter matrix and the filter-covered sector of the layer to produce a single activation level having *shared weights* and a *shared bias* registered in a single *destination pixel* that is positioned in the center of the registered section of the newly *convolved layer*. The weights and bias sharing scheme greatly reduce the computational burden and may also reduce noise because of the convolution of all the shared weights and biases of the convolved area to a single destination pixel.

The process for one filter acting on an input matrix layer with the row vector inner product computation to produce a destination pixel on a new convolved layer is shown schematically in Figure 13.4.[3]

Many different filters may be employed one after the other to produce multiple individual feature maps, and the feature maps can be combined by producing the destination pixels in new convolved layers one by one, producing a final convolved layer feature map matrix.

A convolutional neural network employs convolutional filters to delineate and position prominent features and to retrocede specious, inconsequential features, and noise to produce a series of convolved layers (*inception layers*), using a more refined feature extraction to generate a more accurate image. Inception layers were used for example in Google's cancer tumor diagnosis and in *GoogleNet*'s winning the computer vision *ImageNet Large-Scale Visual Recognition Challenge*.

In supervised learning, a typical training set input might be a two-dimensional 480 × 480 pixel image presented to the CNN's initially "blank" input layer of a random distribution of weights and biases. A 3 × 3 window matrix filter scans the training set input matrix for specific features such as edges, shapes, and colors.

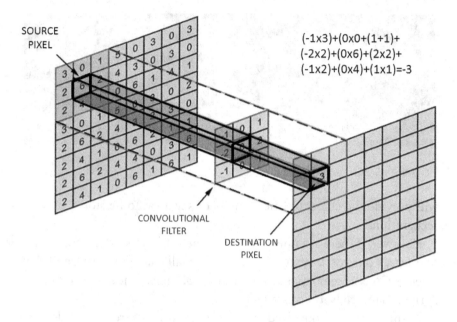

FIGURE 13.4 Convolutional neural network filter operation.

The filter can be composed of a sector of the input layer matrix itself, a specific feature filter such as the *Sobel Gx* edge filter or a Gaussian random distribution of element values that itself can be trained through multiple slides over the input layer matrix and learn from the sharpening or dampening of features. Again, however, different random initialization distributions should be used to avoid the CNN "learning" the initialization and thus biasing the convolutions and the output.

Generally, the first hidden layer detects "high-level" features such as edges from clear shifts in the neuron activation level, for example, the triangular edges of a cat's ears, the second hidden layer detects "low-level features" such as the paws, and succeeding layers extract or refine features like patterns in the fur in a process of *hierarchical feature extraction*.

Conv2D filters are typically used in the first few convolutional layers to extract high-level features and are usually stacked in each convolutional layer. An inception layer convolves different sizes in parallel, from the most precise (1 × 1) to bigger (5 × 5) filters, which extracts detail while covering a larger area. Unlabeled training set data can be *clustered* with extractions of feature similarities.

After extracting the features by parameterization commensurate with minimizing the cost function of the training set data, the combined result

FIGURE 13.5   Cat and dog look-alike.

will provide the characteristic features of say a cat to be stored for later recognition.

However, if there are only a few prominent features extracted, such as four legs, a tail, fur, triangular ears, and a small triangular nose, and classified in CNN's memory as a cat in Figure 13.5 at left, the small dog shown at right may well be classified as a cat.[4]

In this case, the convolved features of the cat require a more detailed feature extraction, and a likely candidate is the cat's pink nose, which may be convolved by a nose-color filter or hand-engineered as a bias.

As the CNN runs through the training sets, just like the artificial neuron layers, the filters themselves can also learn from comparisons with the evolving artificial neuron image. As such, the filters can provide bottom-up feature information that *a priori* may not be known, and thus hand-engineered filters may not be able to extract or locate unusual or unexpected features.

Generally, the array of image input pixels is represented in a computer as a volume matrix with two-dimensional pixels with a third dimension holding the primary colors, typically red, green, and blue (RGB) in proportions that can together generate any color.

The color can be represented by a simple equation,

$$Any\ Color = rR + gG + bB$$

where *r*, *g*, and *b* are the component amounts of each primary color. In a three-dimensional coordinate system with red, green, and blue axes, any color can be represented by a point in the *3D color spectrum* from the value of the *r*, *g*, and *b* factors on the RGB primary colors (Figure 13.6).

However, one of the main difficulties of color recognition is that there are three important independent attributes, *hue*, *intensity*, and *saturation*, that depend critically on the type, angle, and illumination of the object.

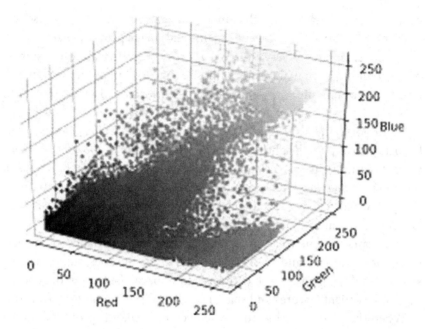

FIGURE 13.6   3D color spectrum.

To recognize hue, computer vision must first determine the intensity and saturation; the intensity can be taken as the average of the three RGB intensity values and the saturation is the ratio of color to illumination; however, the color distinctions in computer vision may vary under different lighting conditions.[5]

Combinations of shape, texture, and color features can be extracted for classification of, for example, strawberries and bananas in a basket of fruits based solely on their different shape and color, but separating strawberries from tomatoes would require size scale and texture analysis.

However, even abrupt changes can be obscured by *noise* in the form of variations in texture, scratches, or electronic instability. Such noise can be *smoothed* by replacing the pixel brightness value with the average or median of itself and its neighbors, thus eliminating the noise but preserving the contrast.

Depth perception is gained by two separated cameras for stereographic imaging, but this requires the correlation of the corresponding points of each image pixel. This can be done by reducing the greyscale arrays to *edge maps* (after extracting edges from the image, map them to form an edge map), scanning the maps to identify the corresponding points, and then measuring the distance to each camera's image plane, and from those differences, reconstructing a three-dimensional image pixel-by-pixel.

This is a seemingly very involved process requiring almost instantaneous multiple computations for almost every pixel in an image, but this is just what a computer can do, in this case almost as well as two biological eyes.

An important feature for recognition is an object's *texture*; that is, regular patterns of pixel brightness, such as the *structural analysis* of *tokens* (salient features) like kernels in an ear of corn and *statistical analysis of directional coherence* like a cat's fur. Tokens and coherence can be relatively easy to detect by the statistical probability that a pixel's intensity will be similar to that of its near neighbors.

The translational, rotational, crops, and flips invariance of objects (move or rotate a cat and it's still a cat), and different poses, such as sitting serenely and lying on its back with legs akimbo, can be clustered, depending on the salient features and spatial relationships among those features.

When the hierarchical feature map is completed, it is projected onto a final convolutional layer which is *flattened* (the matrix is *vectorized*) into a *fully connected* decisional column vector. Flattening a matrix is achieved by successively stacking the matrix column vector elements end-to-end to form a (very) long column vector.[6]

The fully connected output layer is an N-dimensional column vector, wherein each element represents the probability that the input image is of a certain class, for instance, in the classification of pet images, it will have high values in the feature maps that show round eyes, pink triangular noses, and pointy ears. The column vector displays statistical probabilities, for instance, for the classification of household pets arrayed in the output vector as[7]

$$[\text{cat, parrot, canary, dog, hamster, rabbit}]^{T},$$

and suppose the decision vector elements are

$$[0.6, 0.05, 0.03, 0.2, 0.1, 0.2,]^{T},$$

then the image presented to the CNN is most likely a cat.

Detecting, segmenting, and locating an object in a scene require distinguishing the object through the noise of other objects and the immediate surroundings. In these more complicated information transmissions, using techniques such as *bounding boxes* to isolate, and *rich feature hierarchies* for multiple class recognition, allows more accurate re-construction of the entire scene.[8]

There is of course a trade-off: multiple filters and many convolutional layers greatly increase the size of the artificial neural network and the computational burden. However, all of these calculations for convolutional neural network feature extraction can be performed very efficiently by packaged computational software freely accessible from, among others, the *Python* computer language using the *PyTorch* framework and *TensorFlow* platform.

DCNNs, for example, are used in assembly line manufacturing, medical diagnostics, *Facebook* to tag photos, and of course by self-driving cars, as well as many other uses now, and in the future particularly for robot vision.

HYPERPARAMETERIZATION

A convolutional neural network can be tuned to accelerate convergence and avoid overfitting or underfitting data by utilizing *hyperparameters* such as *learning rate, stride, padding,* and *pooling*; the computational burden can also be reduced by adjusting the resolution and dimension size of the convolutional layers. In practice, experience is often the best guide in choosing hyperparameterizations and combinations thereof.

A *learning rate* η adjusts the speed of gradient descent by specifying the size of the steps of the gradient descent; bigger step sizes produce speedier learning, but too big a step size may skip over the desired minimum.

The transposed (superscript T) gradient vector of the Cost function with respect to a number of m weights w_k,

$$\nabla C = \left[\frac{\partial C}{\partial w_1}, \frac{\partial C}{\partial w_2}, \ldots, \frac{\partial C}{\partial w_m} \right]^T ,$$

if a small change in the variable w_k with respect to the gradient is factored by the learning rate η,

$$\Delta w_k = -\eta \nabla C$$

and the weights are iterated as

$$w_k \rightarrow w_k - \eta \nabla C,$$

this then is just the process of gradient descent with an adjustable learning rate factor η specifying the descent step size. Adjusting the step size and thus the speed of convergence can also be used to *stabilize* the gradient descent of the cost function to avoid slowdown or stoppage. Trial and error is most often used to find the best gradient descent step size for optimal learning.

The *stride* is the step size of the convolutional filter as it slides over the matrix, it can be increased to reduce receptive field overlap and produce faster coverage over the matrix layers while concomitantly reducing computational burden. However, if the stride is too large, the filter may skip over or misinterpret some features.

Since the feature maps are the same size as the filters, they will be smaller than the size of the input layer, so the feature map matrix can be *padded* with zeros around the inner periphery of the matrix to ensure that the filter and stride will successfully register with the convolved layer matrix.

The output size in height/length dimensions of a convolutional layer in terms of these hyperparameters is given by

$$Output\ Size\left(\frac{height}{length}\right) = \frac{(W - K + 2P)}{S} + 1$$

where *Output Size* is the height/length of the layer output matrix, W is the input matrix height/length dimension, K is the filter dimension ratio height/length, P is the padding, and S is the stride.

The choice of hyperparameters and filter dimensions will depend largely on the computer vision task at hand and considerations of computational burden.

With this in mind, a *pooling* (also called *downsampling* and *subsampling*) *layer* reduces the dimensions of a convolved feature by matrix multiplying an input layer matrix by a 2 × 2 pooling matrix of stride 2 which outputs either the maximum value or the average value in the sector that the filter convolves, thereby achieving significant dimensional reduction to reduce the computational burden.

Pooling is based on the idea that if a specific feature is known to be in the input matrix by having a high activation level, its exact position is not as significant as its position relative to other features. The resulting down-sampled feature maps are more robust with regard to changes in the position of the feature in the image, so its dimensions can be reduced;

this is called *local translation invariance*. Pooling may also help to extract positionally and rotationally invariant dominant features, reduce noise, and avoid overfitting.

DCNNs can also improve their range of discernment by *augmenting* data to form a broader group by including other image representations while keeping the image label the same. By emulation of the biological brain's *parietal lobe* for recognition of rotational and translational invariance, segment identification and association group expansion, together with simple image pixel shifting, horizontal and vertical flips, random crops, color jitters, translations, rotations, and so on, the original feature characteristic is maintained, but the class is expanded several-fold to include variations of the same, without the need for fresh data.

If a trained network with already tuned parameters is concatenated to a new network with relevant objectives, the new network can "fine-tune" the pre-trained network with only *relevant* new data that is fed to the new network to enhance a specific recognition capability.

This so-called *transfer learning* is implemented by freezing all the gradient descent parameters of all the layers of the trained network, removing the fully connected layer and replacing it with the input layer of the new training network, and then proceeding with the training of the new network with data more relevant to the task at hand. In this way, the features already extracted by the *pre-trained network* do not have to be newly identified by the concatenated network, they just are transferred to the new network and the concatenated network can be taught new, more specific, and more subtle data.[9]

NOTES

1. Vision system image various generic parts, eye, and brain from cs231n .github.io AI course in the public domain.
2. Premier players like Kobe Bryant and LeBron James with relatively flat jump shots are so talented that their percentage is still pretty good, but the high percentages of high-arcing 3-point shooters like Steve Nash, Ray Allen, Stephen Curry, and the improbable Steve Kerr (whose very high shooting arcs resulted in his record for the highest percentage of 3-point shots made in the NBA) are proof of the efficacy of the dot product for long-range shooting in basketball. Incidentally, the backspin of a shot helps to keep the ball on the desired trajectory, so Klay Thompson's high-arcing, tight backspin 3-pointers are not only beautiful to watch but designed for maximum probability of a score.

3. Figure and further explanation is available from the excellent article by Conelisse, C., 2019, *An Intuitive Guide to Convolutional Neural Networks*, freecodecamp.org.
4. Image from Stanford University, Computer Science, CS 231n_02 Spring, personal communication, and available at Adeshpand3.github.io
5. For details on color reproduction, see *Time-Life* 1989 series *Understanding Computers*, and the author's book, Chen, R.H., 2011, *Liquid Crystal Displays, Fundamental Physics and Technology*, Wiley.
6. A matrix may be flattened into a column matrix automatically by, for example, numpy.matrix.flatten.
7. The output layer neurons are arrayed in a column vector, which is conveniently represented in text by its row vector transpose $[0.8]^T$, with the superscript T denoting the transpose of matrix elements from row to column and vice versa.
8. Rich feature hierarchies combine multiple low-level image features with high-level context and with *semantic segmentation* associates a label or categorization with every pixel in an image to classify and recognize objects, such as an autonomous car must recognize other cars, pedestrians, red light, etc. There are many other methods for detection, segmentation, and position of objects, including for example, RCNN, Fast RCNN, Faster RCNN, MultiBox, Bayesian Optimization,Multi-region, RCNN Minus R, Image Windows, Semantic Seg, Unconstrained Video, Shape Guided, Object Regions, and Shape Sharing, See Adeshpande3.github.io, R. Girshick, *Rich Feature Hierarchies for Accurate Object Detection and Semantic Segmentation*, Tech Report (vol. 5), arXiv 1311.2542v5[cs.CV], 22Oct 2014.
9. For example, J. Yosinski, *et al.*, "How transferable are features in deep neural networks?", arXiv.1411.1792v1[cs.LG] 6Nov2-14.

CHAPTER 14

Imagenet and Model Fitting

T HE DEEP CONVOLUTIONAL NEURAL network (DCNN), after training on a database, could extract features of objects, but the utility of the extraction lay in the identification of the object; therefore, for general use, the DCNN's database must be comprehensively organized for general recognition purposes.

Even a simple inanimate object like the common hammer can be separated into claw, ball pein, cross pein, straight pein, pin, club, mallet, soft-faced, nail-punch woodcarver, upholstery, and sledge-, bench-, power-, and spring-hammers, all belonging to the genus *hammer*. Humans furthermore can identify the hammer from only a small segment appearing in a toolbox, for instance, the edge of the claw, and from the size of the toolbox distinguish the claw hammer from a crowbar.

So the identification of an object differs not only in appearance but also in utility and its environment. From this, one with a mathematical bent might think of classifying in terms of groups, using the mathematical theories of sets and subsets.

It so happened that in 1985, George Miller at Princeton produced the English lexical (vocabulary word) database *WordNet*, which grouped nouns, verbs, adjectives, and adverbs into sets of cognitive synonym-sets (*synsets*) and subsets that are related to meaning as a subset of a larger set and is given a name, part-of-speech, and an index number.

A WordNet noun hierarchy is like a biological taxonomy, for instance, a cat, taking a superset as Kingdom (*animalia*) and subsets Phylum

DOI: 10.1201/9781003463542-14

(chordata), Class (mammalia), Order (*carnivore*), Family (*felidae*), Genus (*feli*), and Species (*catus*). A dog taxonomy is *animalia, chordata, mammalia, carnivore, canidae, canis,* and *lupus,* respectively.[1]

It can be seen that there are similarities within a taxonomy class, for example, cats and dogs belong to the same Kingdom, Phylum, Class, and Order but diverge thereafter.

Within each larger set (*hypernyms*) are nouns with some common meanings and in each subset has a finer-grained relational noun called *hyponyms*; in this example, at the Species *catus* level, comprises many different kinds of cats (*The International Cat Association* determined that there are 73 kinds of pedigreed cats), all of whom belong to the *synset catus,* and *lupus* has even more kinds of dogs (it is estimated that there are 450 breeds, the most of any mammal).

Words are organized by semantic/lexical relations to distinguish their different *senses* and then identified in a particular sense. The database provides the domain (topic, region, or usage) and then the syntactic categories of the *synsets.*[2]

DCNNs can recognize features; however, the identification of the object or scene requires a comprehensive training set that includes the labeled image and associated annotations of features, and just as humans learn, in place of experience, the DCNN learns and identifies from its labeled training dataset.

Princeton's Li Fei-Fei realized that for a machine to be able to identify and classify *something*, it would first have to have a database of almost *everything*; that is, a very large training and reference set with subsets organized in a hierarchy of images of objects and scenes all labeled and associated in different contexts.

Professor Li took the idea of *WordNet* nouns to categorize sets and subsets of objects and scenes for an image recognition database. She collected images from Google, Yahoo, Flickr, and other websites, organizing them into the matrices of an artificial neural network to learn and apply.

The collection of images and annotations was begun by Princeton students for $10/hour, but Professor Li quickly realized that even a rudimentary collection of images would take at least 90 years and several million dollars to complete.

It was a grand undertaking, but for want of a means of implementation, it was about to be abandoned. Then just in time, a graduate student told her about the Amazon Web Services (AWS) *Mechanical Turk* which beginning in 2005 recruited netizens from all over the world to perform rather menial tasks on their own computers at home for relatively meager payment.[3]

The collection was easily scalable, so that over 50,000 netizens from 167 countries participated, and her database was sufficient and could easily grow larger. Eventually, more than 100 million images were collected and sorted, including 62,000 images of cats alone.

Image searching included *bounding boxes* around the visible part of the object in question in a scene to pick out the object from many other objects in the same scene and thus capable of discerning specific objects in a conglomeration of objects.

After two and a half years of collecting images, *ImageNet* became a very large database of images with annotations and an index that could be downloaded from the Cloud and used for identification, testing, benchmarking, and competitions,

Object identification by DCNNs with reference to ImageNet images can be used, for instance, in self-driving cars, which of course must identify objects such as other cars, pedestrians, traffic signs, and so on.

In 2009, however, her research report talk application at the *Computer Vision and Pattern Recognition* Conference (CVPR) was rejected, and ImageNet was relegated to a humble poster presentation in a corner of the convention hall.[4]

Notwithstanding the dinosaur reviewers, there were those who saw the light, Stanford University offered her a faculty position and Director of the AI Lab, and she was further appointed as the Chief Scientist at Google Cloud in Silicon Valley.

Professor Li promoted ImageNet as a benchmark test for computer vision by proposing the idea of a contest for the accuracy of computer vision machine recognition of 1.2 million *ImageNet* images drawn from 1,000 different categories in the *ImageNet Large-Scale Visual Recognition Challenge* (ILSVRC).

In 2012, *GoogLeNet*'s DCNN, split into two parts and partitioned across two GPUs, won the ILSVRC, and two years later, its 22-layer, 9-inception module DCNN with 94% accuracy defeated not only the competitor machines but also routinely performed better than human beings at image recognition.[5]

This amount of data and care of collecting, classifying, and annotating brings to mind Charles Darwin, who did just that for the Earth's geological formations, flora, and fauna.

On the toll of decades of observing, analyzing, collecting specimens, classifying, generalizing, and recording, he ultimately formulated his epochal theory, but in his *Autobiography* lamented,[6]

> *My mind seems to have become a kind of machine for grinding general laws out of large collections of facts such that neither music or literature nor appreciation of fine scenery held any pleasure any longer.*

Indeed, Darwin was the ultimate human manifestation of a deep convolutional neural network algorithm for searching, classifying, and generalizing huge amounts of data which he duly organized and recorded.

As ImageNet and the recognition capability of DCNNs grew, such a machine can do all the above without want of human pleasures, so life sciences academia can look forward to many Darwinian robots tirelessly performing natural science in geology, flora, and fauna, and indeed any observational pursuit.

Darwin's classic 1859 book *Origin of the Species* set forth the induction, deduction, and generalization that established the Theory of Evolution, one of the greatest scientific achievements of mankind.

Darwin first established the branching patterns of flora and fauna evolution based on his theory of natural selection; he later turned to human evolution in two books, *The Descent of Man* (1871) and *The Expression of Emotion in Man and Animals* (1872). He thus amalgamated the classification of biological objects and their emotive expression (the "emotional knowledge" of Chapter 32?), which could be used by DCNNs and ImageNet for finer detail of objects, including the controversial detection of emotion from computer vision facial recognition (for example, wider open eyes for surprise and open mouth betraying fear).[7]

Particular traits of different nationalities, either innate or cultured, for example relatively reserved Germans and Japanese, compared to more expansive Italians and Nigerians could also form a class for deeper and more distinguishing recognition (although fraught with the risk of stereotyping). Woe to the Italian negotiator who believes she has closed the deal with a Japanese company because its representatives all nodded in seeming assent to her proposal, their response actually being no more than a polite acknowledgment that they had heard what she was saying.

The enormity of variations of facial expression, body language, hand gestures, national and local customs, and so on in diverse cultures would require huge additional amounts of extremely detailed data, yet still fraught with indistinct subtlety, ambiguity, context, and subject to change with the substantive events of the times and the subject's physical and emotional state at the time of observation, all the while skirting the issues of privacy and prejudice.

MODEL FITTING

The functional operations model for a thinking machine can be represented by a simple mapping function,

$$y = f(x)$$

where given the stimulus input variables x, the model $f(x)$ maps them onto the output y, the response.

In artificial neural networks, the mapping function $f(x)$ is constructed by training the network on a set of labeled data, that is, by *induction* from the *abstracting of common characteristics* found in the labeled training set data to form a model *generalization* for recognizing newly presented data by *deduction*, going from the model to a particular case.

In supervised learning, if the output y does not accurately match the labeled input data x, the difference between the output and the labeled training data (the *Cost*) will be minimized and the adjusted parameterized data fed back into the algorithm model as shown in Figure 14.1.

If after many iterated runs of Cost minimization, if the parameterizing algorithm does not converge to the training set data, then the algorithm model itself may require a *system* calibration or a complete change.

After successful supervised training, the calibrated model presumably has learned the common characteristics of the training set data and is able to generalize those characteristics to recognize new images and probability distributions or perform classification on new unstructured and unlabeled data or images.

A mathematical definition of *generalization* is,[8]

> *There exists a set of elements that possesses common characteristics shared by those elements sufficient to form a conceptual model that can perform deductive inferences.*

In other words, the training set data must have sufficient common characteristics to configure the model so its deduced generalizations can classify

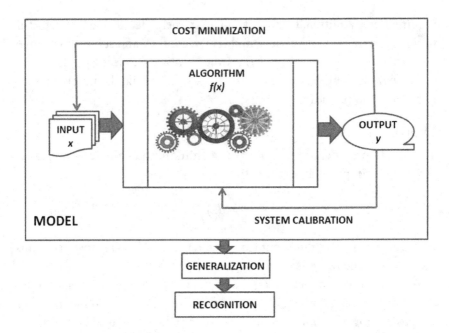

FIGURE 14.1   Algorithm model.

new data as belonging to some learned set of data or predict the consequences that any new data implies.

How well the model $f(x)$ can classify and predict is determined by the *goodness of fit* of the model on the new data. The model *underfits* when it cannot adequately capture the abstract common characteristics of the training data and simply cannot generalize anything from the training data. Underfitting is easy to detect from the model's poor performance on the training set data commonly caused by *under-training* the model, and the remedy is simply to provide more training set data and more runs (*epochs*) through that data.

An analogy is the lazy schoolboy who doesn't do enough of the reading assignments and does not learn the basic principles of the course subject and thus performs poorly on examinations.

The model can also *overfit*, a definition from the Oxford Dictionary is,[9]

> *The production of an analysis that corresponds too closely to a particular set of data, and may therefore fail to fit new data or predict future observations reliably.*

The machine learning model overfits when the model unwittingly incorporates irrelevant detail or noise in the training set data into its generalization as if they were part of the essential common characteristics of the training data.

Overfitting means that the model has been *over-trained* and like the diligent but dull schoolboy who tirelessly memorizes the math problem solutions in the textbook instead of learning the abstract generalizations from the problems and their solutions to apply them to new problems in an examination.

Possible reasons for overfitting are a decision tree that has too many branches or a neural network with too many hidden layers irrelevant to the desired generalized abstraction, amounting to incidental noise, so much so that in the jumble of extraneous information, it is unable to abstractly deduce recognition of an input of new data germane to the what the training set wants to generalize.

A good fit can be developed by *resampling*, for instance *k-fold cross-validation*, which is just a fancy way of saying take k different subsets of the training dataset and leave one subset, train each of the $(k-1)$ subsets, and take the average of performance metrics, such as for classification, the ratio of correct predictions and all predictions, mean square error (MSE), root MSE, sensitivity, and so on; or use the $(k-1)$ subsets to evaluate the accuracy of the trained model.

There are many different modes of validation testing; one is to separate a subset of the training data to use as a *validation dataset*, run the model on the rest of the data, and note the results. Then run the validation dataset to determine any differences. Another validation test of new data can be run on different models and compare the performance of each model.

This is like the clever schoolgirl who gleans from the quizzes what will be on the final examination. It is usually used to tune hyperparameters and has become an essential tool for improving machine learning algorithm accuracy, but it cannot be used too much as the machine may learn the validation dataset instead of the training dataset.

In another test, a *test dataset* can be run that is independent of the training dataset but follows the same probability distribution. If the results are similar to the training dataset run, that is an indication that the model does not overfit. This is the case of the brilliant schoolgirl who has learned how to generalize all the subject matter of a course, thereby understanding the essence of the subject.

Finding the optimum learning regime can be achieved by observing the rate of improvement of the model in accurately recognizing the labeled training set data over the training *epochs*, and when the rate of improvement approaches zero, the training should be stopped to avoid overfitting. This is like the able schoolteacher who avoids boring the students by repeating the same material.

Underfitting and overfitting are the twin gremlins plaguing artificial neural networks; fortunately, in addition to test and validation datasets, there are many ways of dealing with them in machine learning modeling.

Different weights and biases parameter initializations can be employed, such as Bayesian, *Gaussian Mixed Model*, and *Factor Analysis*, giving the artificial neural network a head-start in banishing the gremlins.[10]

Model underfitting and particularly overfitting both also can be alleviated by *regularization* methods such as the hyperparameters *L1* and *L2*, *dropout*, and *artificial expansion* of the training data.

Too few neurons in the hidden layers may result in feature maps extracted from the data that miss significant characteristics, seriously underfitting the input data. However, on the other hand, using too many neurons in the hidden layers can result in activating irrelevant data and noise, seriously overfitting the data. In this case, the network has so much processing capacity that the data in the training set is too limited to train all the neurons in the hidden layers, and the neurons "find" extraneous data.

The overfitting of training set data by excess neurons can be alleviated, simply enough, by *dropping out* a randomly chosen set of activations in a layer by setting them all to zero. Different sets of neurons can be dropped out, so it is like training a different neural network after each dropout.[11]

This is therefore a test of network perspicuity; if the model then determines the common characteristics regardless of the dropped-out activations, it is like the brilliant student who gets right to the heart of the subject knowing what is relevant and avoids going in too deeply, thereby avoiding overfitting.

As one of the pioneers of modern AI and a recipient of the 2018 Turing Award Yann LeCun explained,

> *This technique reduces complex co-adaptations of neurons, since a neuron cannot rely on the presence of particular other neurons. It is, therefore, forced to learn more robust features that are useful in conjunction with many different random subsets of the other neurons.*

On the other hand, artificially *expanding* the training data can make the artificial neural network model more precise, for example in improved speech recognition, adding background noise to *best-fit* real-life listening situations.

Overfitting can also be ameliorated by hyperparameter *regularization* techniques, such as *weight decay (L1 regularization)*, which adds a term to the Cost function, thereby stabilizing its gradient descent, proving useful when different runs of the artificial neural network produce quite different results. In *L2 regularization*, the Cost function is modified by adding the sum of the absolute values of the weights for stabilization of its gradient descent backpropagation.

NOTES

1. These are one of the various different classification and names in taxonomy.
2. WordNet is licensed free by BSD.
3. Today, the payment is $6 per task, amounting to less than $2 per unit. It has been estimated that full-time Mechanical Turks can earn an average of $46,000/year.
4. Professor Li had difficulty obtaining research grants for the ImageNet project, with comments such as "it was shameful that Princeton would research this topic" and "the only strength of the proposal was that Li was a woman". Gershgom, D., *The data that transformed AI research, and possibly the world*, Quartz, qz.com.
5. *ImageNet* is a large-scale hierarchical image database designed by Jia Deng, Wei Dong, Richard Socher, Li-Jia Li, Kai Li, and Li Fei-Fei in 2009; Alex Krizhevsky, Ilya Sutskever, and Geoffrey E. Hinton of the University of Toronto won the 2010 ILSVRC. Two CPUs were necessary because the *Nvidia GeForce GTX 580* had insufficient on-chip memory.
6. *The Autobiography of Charles Darwin*, John Murray III publisher (1887).
7. Refer to the definitions of "intelligence" in Chapter 32, which included the perplexing "emotional knowledge" as a measure of "intelligence".
8. This is the author's definition, following many perhaps better formal descriptions.
9. Author changed the Oxford definition slightly to reflect modern machine learning usage. For some reason, Oxford does not define "underfitting".
10. These will be discussed in the following chapters.
11. See N. Svrisastava, G. Hinton, *et al.* 2014, *A Simple Way to Prevent Neural Networks from Overfitting*, J. Machine Learning Res, p. 15.

CHAPTER 15

Markov Chain Monte Carlo Simulation

M ATHEMATICAL MODELS OF MACROSCOPIC physical phenomena and engineering systems almost always take the form of second-order partial differential mostly because Newton's second law is a second-order differential equation. For deterministic physical systems, those equations are mostly *linear*, meaning that the coefficients of the derivative terms were constants or functions only of an independent variable (and not a dependent variable), and there were no derivatives multiplying each other or themselves squared, cubed, or raised to higher powers.

For relatively simple systems, linear differential equations often could be solved in closed form, meaning in terms of the elementary functions of polynomials, sine, cosine, exponentials, natural logs, and combinations thereof, and these solutions would fully describe the physical situation under different initial and boundary conditions, thereby allowing mathematical description and prediction.[1]

However, most physical systems in real life are almost always non-linear and thereby not solvable in closed form, so using the *finite-differences* first of tiny mechanical turns of cogwheels of the early *differential analyzers* and later digitally incrementing the independent variable and correspondingly differentiating the dependent variable in response to the increments of the independent variable; the equations were solved numerically at first very slowly by young women and adding machines and later very quickly by numerical analysis programming using digital computers, and taking derivatives to hasten convergence (Newton's method).

 DOI: 10.1201/9781003463542-15

The complexities of real-world situations, however, involved many factors represented by multiple components and terms in the differential equations, many independent and dependent variables, and all manner of parameters and probabilities, each solution also dependent on different initial and boundary conditions.

Newtonian mechanics furthermore does not hold in the microscopic world of molecules, atoms, and nuclei, where the quantum mechanical second-order differential equations of Schödinger and the matrices of Heisenberg dealt with the Born probabilities of wavefunctions, rather than the determinism of a Newtonian macroscopic calculation.[2]

THE ATOMIC BOMB

The Atomic Bomb releases energy because of the splitting of the nucleus of radioactive uranium-235 (^{235}U) or plutonium ^{239}Pu and ^{241}Pu, and the masses of the fission fragments are less than the mass of the original nucleus.

The *mass defect*, according to Einstein's iconic mass-energy equivalence equation $E = mc^2$, is equivalent to a release of the *binding energy* of the protons and neutrons in the original ^{235}U nuclei.[3]

A uranium Atomic Bomb explodes when a nucleus is split and protons and neutrons of ^{235}U are released, and although the positively charged protons tend to avoid collisions with other protons, the neutral neutrons will form a neutron flux and collide with other ^{235}U nuclei, instigating further nuclear fission and again releasing copious protons and neutrons that increase the density of the neutron flux.

When the nuclear fission *effective neutron multiplication factor* $k = 1$ and the amount of ^{235}U has reached a *critical mass*, a spontaneous chain reaction of fission reactions suddenly releases all the proton- and neutron-binding energies in a nuclear explosion.

For example, the ^{235}U Atomic Bomb dropped on Hiroshima absorbed a neutron and gained mass, splitting into ^{93}Kr and ^{141}Ba and emitting γ radiation and three neutrons in the fission.

In designing the A-Bomb, scientists first tried to use differential and integral equations to deterministically follow the neutrons in their collisions with the uranium nuclei and the subsequent release of more protons, neutrons, and radiation. However, the spreading cascade of collisions could produce more than a billion more neutrons, and the *neutron flux* could go on to collide with billions of other uranium nuclei, and the collisions would produce even more billions of neutrons in ever-spreading branches of a chain reaction tree of subsequent collisions.

Furthermore, each collision, fission, and release of neutrons event was quantum mechanical and thus could not be absolutely determined, only the probabilities of collision events were known.

The myriad possible interactions in fission reactions and neutron collisions depend on the probabilities of the particles' *elastic scattering* (change direction but not energy), *inelastic collisions* with nuclei (change of direction and energy), *absorption* by the nuclei, and possible *fission* of the nuclei.

These reactions and their probabilities constituted very complicated non-linear, second-order differential equations that were clearly not amenable to a deterministic kinematic (study of motion) calculation of each and every atom nucleus, proton, and neutron, and all of the processes were further subject to the initial conditions in the reaction chamber and the boundary conditions of the bomb casing.

SOLITAIRE

One of those participating in the development of the Atomic Bomb was the mathematician Stanislaw Ulam, an immigrant from Poland threatened by the Nazis in 1935; he joined the Los Alamos team in 1943, taking up work on analyzing the neutron flux problem, but with no more success in the kinematics study than others.

The uranium critical mass and neutron flux problems stuck with Ulam, however, and in 1945 when struck with viral encephalitis, he was confined to hospital with nothing to do but play solitaire all day, which he did, over and over again.

Thinking like the mathematician that he was, Ulam wondered if there was any statistical way he could find keys to successfully complete each game, so he began to record the sequence of cards and their play and the game's final layout, attempting to find some probabilities in a pattern of play that would lead to successfully placing all the cards. He believed that if he could just observe a large enough number of games, he would be able to discover those patterns and model a successful card layout.

In a fair shuffle, the sequence of cards drawn from a 52-card deck will always be random, and the cards can be arranged in more ways than the number of atoms in our galaxy (about 10^{67}); however, there were only a few placement possibilities in the array for a drawn card, so the game could be modeled.

However, a card placement choice would not depend on any preceding layout state of the cards, and each card placement would change the layout state, and each layout would be different for each game, so an astronomical number of games would be required to discover the probability of success.

A stochastic ensemble of a sequence of random variables X_i (cards) belonging to a finite configuration space (card layout) of separated nodes connected by chains (card placement possibilities) among which an agent can choose to take but with random probabilities P_{ij} of steps towards other nodes (card placement probabilities), whereby the probability distribution of states in the process depends only on the present state (card layout).

Ulam knew that a Markov chain with the *transition matrix* of probabilities P, where each element of the P matrix represents the probability of a step from A to B and C, B to A and C, C to B and A, and steps to itself (meaning "wait"), does not depend on any past states (the *Markovian property*), a schematic example of which is shown in Figure 15.1.

The individual elements of the conditional transition matrix probabilities P are given by (k is a dummy index)[4]

$$P_{ij} = Probability\left(X_{k+1} = j \mid X_k = i\right)$$

If there is a positive probability, no matter how low, that an agent can move onto any other nodes in the chain, it is *irreducible*; if the agent cannot go around endlessly in cycles of the same nodes, the chain is *aperiodic*; and if the agent can explore every node, the chain is *erdogic*.[5]

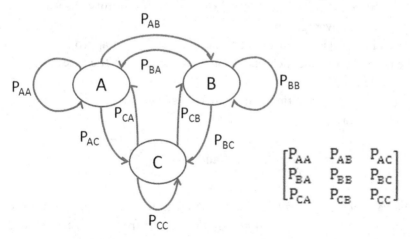

FIGURE 15.1   Markov Chain with transition matrix.

The Markov chain seemed to satisfy the game's requirements; Ulam believed that within the domain of possible steps, there could be a definitive probability of success if all the process probabilities of card arrays are simulated in a large number of games to statistically model the 10^{67} different distributions of a deck of randomly shuffled cards. Further, since there are states where possible placements of the card depend on probabilities that will go towards reaching the objective of laying out all of the cards, the Markov chain could apply to the collision and absorption probabilities of the buildup of the collective mass of ^{235}U and the density of the neutron flux required for an Atomic Bomb explosion.

The random distribution of cards in a shuffled deck could be taken from sampling a random distribution of numbers representing the cards, and then by repeated sampling and aggregating the results, if he could simulate a sufficient number of games, he could determine the probability of success, both for solitaire and Atomic Bomb detonation.

A simulation algorithm employing a random distribution is called "Monte Carlo" after the name of a casino in Las Vegas itself named after the gambling capital of Europe, evincing the gambling industries' propaganda of a completely random chance at winning (something that does not include the assured profits of the casino), based as it is on the *Law of Large Numbers*. Also called *Bernoulli's law*, it is the ultimate regression to the mean of any activity after a very large number of trials and is the foundation of the efficacy of a Monte Carlo simulation of a complex system.[6]

As the size of a statistical sample N approaches infinity, the *variance σ^2*, which as the average squared standard deviations σ, is a measure of the variability of data from the arithmetic mean μ, will approach zero, and the regression to the mean probability will approach the true probability of any activity.

The standard deviation σ measures the dispersion of the data and is just the root-mean-square of the variance,

$$\sigma^2 = \frac{1}{N}\sum_i (x_i - \mu)^2 \text{ and } \sigma = \sqrt{\frac{1}{N}\sum_i (x_i - \mu)^2}$$

The variance as the *square* of the difference will always be positive, and squaring furthermore will highlight the outliers from μ, and regress to the mean (if an absolute value is used instead of squaring, it will regress to the median rather than to the mean).

The variance will approach zero as the number of trials approaches infinity, so in the example of a fair coin toss, the probability of heads or tails will eventually proceed to the mean probability of 0.5, with a variance of zero as the number of trials approaches infinity as shown in Figure 15.2.

A simple example of a system with various sub-systems is a pair of dice that has 36 possible combinations; by continually rolling the dice many times and recording the result, the probabilities of a given combination follow the law of large numbers, so the probability of a "7" is six times higher than the probability of a "2" since there are six ways of the dice adding to 7 and one for a two "1"s. The probability distribution will be a Normal distribution curve with the "7" at the peak and "2" and double six "12" on the wings of the Normal distribution bell curve.

The regression to the mean of any system and its sub-systems after many simulation runs will regress to the most probable outcome, which is the most that the Bomb designers could hope for in establishing the parameters for detonation.

A simplified example of a Markov chain Monte Carlo Bomb explosion simulation (MCMC) is a source of neutrons passing through a cavity containing ^{235}U that may either be scattered or absorbed by the uranium nuclei with the latter possibly resulting in fission depending on the density of ^{235}U nuclei and the accumulating density and energy of the neutron flux. The problem then is one of neutron diffusion and fission multiplication of neutrons, the probabilities of which can be found from nuclear physics experimental results.

COIN TOSS

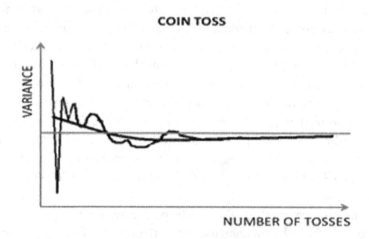

FIGURE 15.2   Coin toss demonstration of law of large numbers.

The mean distance that the neutrons will travel without being scattered or absorbed (*mean free path*) and the probability of a neutron collision with a ^{235}U nucleus (*collision cross-section*) also depend on the kinetic energy of the neutrons.

It is known (and completely believable) that the probability density P of a neutron traveling a distance x before being scattered or absorbed (free path) decreases exponentially depending on the density of nuclei ρ and their collision cross-section ζ, the infinitesimal probability for an infinitesimal travel distance dx thus is given by

$$dP = \rho\sigma e^{-\rho\zeta x}dx$$

Integrating this equation will give the density of the neutron flux (number of neutrons passing per unit square area) as

$$Neutron\ Flux = \frac{Number\ of\ Neutrons\ Passing\ Through}{Number\ of\ Constructed\ Trajectories}$$

Now x_i the free path length for trial i, which is over the open interval $(0,\infty)$, can be represented by a computer-generated sequence of *pseudorandom numbers* ξ_i uniformly distributed in the interval $(0,1)$ by making the transformation,

$$x_i = -\frac{1}{\rho\sigma}\ln\left(1-\xi_i\right)$$

That is, the neutron free path can be expressed by pseudorandom numbers from the computer through this variable transformation. For example, if experimental data for neutron collisions with ^{235}U shows a 0.9 probability of scattering and only 0.1 probability of absorption leading to fission, and if the ξ_i interval $[0,1]$ is segmented into two groups $[0,0.1]$ and $(0.1,1]$ (note carefully the commas and periods), if say the pseudorandom number generated by the computer, for example, is 0.2, then it belongs in the second larger group $(0.1,1]$, meaning that the neutron has been scattered. Repeating for more and more sets of pseudorandom numbers will give better and better approximations for the *Neutron Flux* equation above.

If a neutron passes through, it is given a "score" $s = 1$ and if absorbed $s = 0$, so the probability of contributing to the neutron flux is given by the mean score $\bar{s}$ where the error is measured by the variance.

Performing the same type of transformation and scoring for scattering angle, the directions of the scattered neutrons can also be simulated, although the possibility of neutrons scattered back into the neutron flux must also be considered, which complicates matters and must be considered in more detailed calculations.

For more complex situations considering protons, including different cavity designs, initial conditions, and so on, the situation can be modeled with experimentally derived cross-section and mean free path parameters for different internal cavity characteristics, and entered into the MCMC simulation.[7]

Now after sampling from a Normal probability distribution and after many runs of the MCMC simulation, the vast number of particles in Atomic Bomb neutron diffusion flux (10^{15}–10^{25}) can be modeled by sampling only 10^5–10^8 trajectories, and for critical mass represented by a function $f(X)$ with sequences of random samples X_i to approximate a desired probability function $P(X)$, where $f(X)$ is proportional to $P(X)$.[8]

The desired probability $P(X)$ for critical mass can be thought of as a *probability density* that is reached by $f(X)$ stepping through the Markov chain in steps commensurate with the greater probability (generally towards the peak of a random distribution), iteratively pushing $f(X)$ closer to $P(X)$, and consistent with Bernoulli's law, the MCMC simulation in aggregate will regress to the mean probability of attaining a critical mass of uranium nuclei to produce a neutron diffusion flux sufficient to form a self-sustaining chain reaction and subsequent detonation.

A large number of neutron trajectories are constructed by sampling from the experimental probability parameter distributions as the neutrons travel through the Bomb cavity. The chain reaction therefore depends on the aggregate outcomes of the total set of trajectories.

The random distribution ensures that the actual probability of a chain reaction will be "covered" by the 10^5–10^8 trajectories, and with the ever-increasing computational capabilities of computers, now including supercomputers, the number of trajectories can be increased to better satisfy the law of large numbers.

Running the Monte Carlo simulation over the Atomic Bomb Markov chain many times will drive the variance of outcomes to zero and thereby reveal the mean probability of attainment of the chain reaction for producing the detonation of the bomb.[9]

If the probability is low, then the thermodynamic and Bomb cavity engineering parameters can be adjusted and the simulation run again and again until conditions for *Critical, Super-Critical,* and *Sub-Critical* situations are known.

It is critical to know the conditions for detonation so that the bomb will detonate when desired and *not* in the laboratory, so a simplified plot of Mass/Energy vs. Time must be calculated from the MCMC simulation of neutron beam diffusion.

The MCMC was developed too late for the Atomic Bomb, but beginning in 1951 the development of the Hydrogen Bomb required fission reactions to generate the radiation to trigger the fusion of the hydrogen isotopes deuterium and tritium.

Ulam first ran the Metropolis-Hastings MCMC algorithm for a system of many interacting particles on John von Neuman's MANIAC computer (derived from the Institute for Advanced Studies (IAS) computer at Princeton) and later successfully completed the simulation on the ENIAC computer at the University of Pennsylvania.

The H-Bomb (the "Super") detonated based on the idea of *radiation implosion* from nuclear fission reactions to produce H-Bomb fusion, with controversy over whether it was Edward Teller or Stanislaw Ulam's idea, but Teller's insight regarding Ulam's Markov chain Monte Carlo simulations was spot-on,[10]

> *Take advantage of the statistical mechanics and take ensemble averages instead of following detailed kinematics.*

An analogy with the bombs can be made for a presidential election prediction. Survey samples from the general population are constructed from demographics such as party affiliation, gender, economic class, ethnicity, and so on, providing in effect event probabilities just as the mean free path and collision cross-sections for neutrons and uranium and plutonium isotope nuclei have probabilities for scattering or absorption and subsequent fission.

Aside from nuclear weapons research and the risks inherent in democratic elections, Monte Carlo simulation has been used in the more beneficent areas of weather prediction, turbulent airflow around jet planes, expansion of the Universe, biological systems, ecology, stock market, sales prediction, economic models, and in any operations research problem such as traffic control and airport passenger flow. Monte Carlo simulation has been gainfully employed in almost all science and engineering, both natural and social, and particularly in simulations of the results of the Big

Data, LLMs, and GPTs (very many numbers and parameters) of artificial intelligence learning, performance, and prediction.

NOTES

1. A differential equation relates a function with its derivatives (rate of change with respect to independent variables) so that its solution can predict outcomes subject to initial and boundary conditions. The differential equations are almost always second-order because the second derivative of position, acceleration, is proportional to the force under Newton's second law, and it is the force that drives many macroscopic physical systems of interest. Examples are the non-linear ballistic projectile differential equations in Chapter 9.

2. Refer to Chapter 27 for some quantum mechanics.

3. Uranium occurs naturally as 99.3% ^{235}U so the 0.7% ^{235}U isotope has to be purified, for example, by high-speed centrifuges that spin out the ^{238}U leaving pure ^{235}U used in the "Little Boy" gun-type triggered Bomb on Hiroshima. A nuclear reactor can breed plutonium from uranium fission, a plutonium "Fat Man" implosion-triggered a ^{239}Pu Bomb which was dropped on Nagasaki.

4. The straight vertical line "|" means a conditional probability, $P(A|B)$ where the probability of A occurring depends on the occurrence of event B, or "given B, the probability of event A" happening.

5. For a deeper mathematical description of a Markov chain, refer to M. Richey 2011, *The Evolution of Markov Chain Monte Carlo Methods*, American Mathematical Monthly, Vol. 117, No. 5, online.

6. Casinos make 65–80% of their profits from slot machines based on the low probability of a random combination of symbols that is a reward much less a jackpot, so from the law of large numbers, one must play a huge number of times before the average is reached, and although some players will win some, they will not play long enough, and lose more before they reach any jackpot. In card games, high rollers betting big money also will not play long enough to regress to a far-off mean winnings, and the zero and double-zero of roulette gives the House a 5.26% edge. Ref. *Finance Monthly*, "Here's how casinos make money".

7. For more detail on MCMC, see J.S. Hendricks 1994, *A Monte Carlo Code for Particle Transport*, Los Alamos Science, Number 2.

8. The $f(x) \propto P(x)$ proportionality is sufficient for the calculation obviating the difficulty to determine a normalization factor.

9. Of course, the Manhattan Project ultimately produced a sound design based on the nuclear physics of the time, parameter testing, analog computing, macroscopic temperature, pressure, and density requirements of the neutron gas flux and the volume and shape of the bomb cavity, all practically engineered by step-by-step experimentation.

10. Ulam was a principal scientist in the development of the Hydrogen Bomb; the crucial idea of radiation pressure from a fission explosion to ignite

fusion has been attributed to either him or the "father of the H-Bomb" Edward Teller, or both or others in an ongoing controversy kept alive by the top-secret classification of H-Bomb development documents. Ref. author's book, Chen, R.H., 2017, *Einstein's Relativity, the Special and General Theories with their Cosmology,* McGraw-Hill Education; R. Rhodes 1986, *The Making of the Atomic Bomb,* Simon & Schuster; Rhodes, R., 1995, *Dark Sun, the Making of the Hydrogen Bomb,* Simon & Schuster for the history of the Bombs. Quote attributed to Teller by Marshall Rosenbluth shortly before Rosenbluth's death in a 2003 presentation "Genesis of Monte Carlo Algorithm for Statistical Mechanics" at the Los Alamos National Laboratory.

CHAPTER 16

Reinforcement Learning

R EINFORCEMENT LEARNING IS PERFORMED by everybody, every day. It requires a situation, a *state* presented to the agent, an *action* responsive to that state, and a reward or punishment for the action. A simple example is a state of a thrown ball, and action by Fido to chase it down and bring it back, and a doggy treat for doing so, or harsh words for either not chasing it down or not bringing it back.

Apart from the rather serious contests of human mental acuity like chess, it seems that since video games were created on computers, it is reasonable to think that silicon-based video game machines (although having no thumbs) could also learn to play them and play them well enough to defeat carbon-based human gamers in what is a decidedly fast-paced reflexive contest of manipulative skill, planning, and intelligence.

The University of Toronto and the company *DeepMind* together developed the DeepMind Gamer which could at once play and learn *how* to play from the rewards and punishments of *reinforcement learning* (RL), something entirely in tune with everyday life of education, work, self-improvement, raising children, and, particularly among the young, video games.

But RL is also used in many disparate professional disciplines such as economics, game theory, industrial automatic control, information theory, operations research, and animal behavior.

The Toronto/DeepMind Video Gamer's very public defeat of expert video gamers without even knowing the rules of the game beforehand was an eye-opening event of artificial intelligence prowess and promise.

Machines employing reinforcement learning were successful in game-playing against humans because the machine learned just like human

DOI: 10.1201/9781003463542-16

beings learn, from the experience of actions resulting in rewards and punishments, and any gamer knows that the more games you play, the better you get.

But unlike humans, a machine's skill can be tirelessly honed through millions of games against not only expert humans but other game-playing machines, and once establishing its superiority, it can self-supervised play against earlier versions of itself, going far beyond the skill of the best humans.

THE MARKOV DECISION PROCESS

A Markov Decision Process (MDP) is a Markov Chain that allows more choices and rewards or punishments for actions. In a video game, the agent player is presented with a state and makes a decision about what to do about it (for example, a wall obstacle); the action chosen by the agent (jump over or go around) will change the state, and given the new, changed state, the MDP will give a reward (or punishment, both called a "reward") based on the agent's playing *policy*.

The agent then will be faced with a new state resulting from this action, and again the MDP will evaluate the new state depending on the agent's policy and give a reward. This continues until the goal of the policy is reached, and the agent has thereby *learned* from the rewards how best to reach the goal.

Reinforcement learning operates as a simple MDP feedback loop operating in successive time increments, as shown in Figure 16.1.

Here the subscripts denote time t and its increment $t + 1$. Reinforcement learning thus constitutes a sequence of state-action (s_t, a_t) pairs that are performed respective to rewards (high positive values) and punishments (low or negative values) for r_t and r_{t+1}). The agent seeks to select actions that maximize the sum of rewards over time.

The Markov Decision Process is used to find an optimal *policy* in decision-making where the outcomes are partly random and partly under

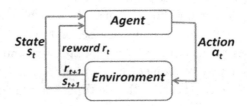

FIGURE 16.1   Reinforcement learning feedback loop.

the control of the agent. Figure 16.2 is a schematic of an example of the process.[1]

At each time step, the process is in some state s_t and the agent can choose any action a_t that emanates from the node edges of s_t; MDP then randomly moves into a new state s_{t+1} and gives a reward $R_a(s_t, s_{t+1})$ based on the agent's policy. The numbers on the arrows are the random probabilities of leading to the next node selected, and the wriggly arrows at the top and bottom are the rewards for a particular choice, as determined by how the choice leads to the agent's policy (say +5 high positive reward; −1 low negative reward).

The so-called *Q-learning algorithm* finds the value of an action in a particular state by the MDP, which over many simulation runs provides an *optimal policy*. Each action is represented by a Q-function that updates the state-action pairs (s_t, a_t) commensurate with the present and subsequent action rewards. Then the highest combination of immediate reward with all possible future rewards gained by later actions a_t is determined using the Bellman equation iteration update with the weighted average of the old value and the new information,

$$Q^{new}\left(s_t,a_t\right) \leftarrow Q\left(s_t,a_t\right) + \alpha \left[r_t + \gamma \max_a Q\left(s_{t+1},a\right) - Q\left(s_t,a_t\right)\right]$$

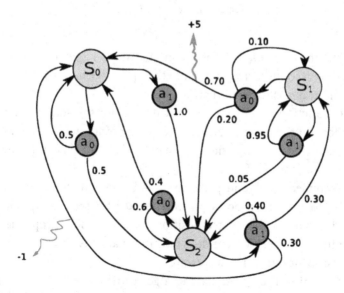

FIGURE 16.2 Markov Decision Process.

where $Q^{new}(s_t,a_t)$ is the new Q value, α is the learning rate, the expression in square brackets is the learned new temporal value wherein r_t is the reward, γ is the *discount factor* in the interval $[0,1]$ that gauges the relative importance of immediate rewards versus future rewards, and $\max Q(s_{t+1},a)$ is the estimate of maximum future value.

The Q-function is initialized by an arbitrary value (thus the playing agent does not need to know the rules of the game), then as time t progresses, the agent selects an action a_t, notes the reward r_t, and enters a new state s_{t+1}, and Q is iteratively updated to Q^{new} to produce the action value for the new state, which value should ideally steadily increase through reinforcement learning rewards, and gradually from simulations the agent iteratively learns how best to play the game to win, that is, learning the optimal policy.

The Q-function can consider delayed rewards through the discount factor γ introduced in later time steps in the game sequence and can act recursively through program *nesting* to encompass those rewards in the algorithm computation.[2]

The goal of the agent is to maximize the total reward. It does this by adding the maximum reward attainable from future states to the reward for achieving its current state, effectively influencing the current action by the potential future reward. This *Goal* is a weighted sum of expected values of the rewards of all future steps, expressed by

$$Goal = max \sum_{t=0}^{\infty} \gamma^t r\left(s_t,a_t\right)$$

This is just the *maximization* of a sum of rewards $r(s_t,a_t)$ over time t multiplied by a discount factor γ raised to the t power, where s_t is the state at a given time and a_t is the action at that time.

The Q-function provides a numerical score for the action taken based on its effect on the game environment by mapping each state-action pair to a number that is determined by the rewards that best contribute to the states reaching the *Goal*.

For example, a pick-and-place robot is being trained by an RL agent controller to give the robot a positive reward for picking up the object and placing it in the designated position, but if the robot drops the object, places it in the wrong place, or does nothing at all, it is given a low or negative punishment number as a "reward".

After running the game-playing algorithm many times in training, the Q-function selects the (s_t,a_t) pair with the highest Q *value* from the game-playing experience. Using feedback from the experience, a scalar reward is sent back for each new action.

Since it has a value between 0 and 1, raising γ to the time t power means that if γ is small, as t increases, the reward is multiplied by a fast-decreasing factor of γ^t and the value of the *Goal* decreases very quickly, so as $\gamma \to 0$ the reward is "myopic" (nearsighted) in the sense that the immediate goals are more important to reaching the *Goal* (for example, in returning a shot in ping-pong).

On the other hand, a larger γ does not reduce the value of the *Goal* so rapidly because as $\gamma \to 1$, the reward is "hyperopic" (farsighted) in that the reward multiplied by a slower-decreasing factor γ^t thus maintains a higher value longer, thereby attaching more relative importance to longer-term goals (for example, in developing pawn structure in chess, *sente* in Go, and more goals in football).

In operation, γ also mathematically prevents the summation in the *Goal* equation from exploding to ∞ and hanging up the computation. The discount factor γ can be hand engineered or machine learned to maximize the *Goal*.

In complex environments, selecting the best action among many choices commensurate with a given state requires the *ranking* of the quality of actions, which is based on a measure of the *value* of (s_t, a_t) pairs; that is, how much do they further the positive accumulation of rewards. A *policy function* π based on those values maps a state s_t to the best-known action a_t,

$$a_t = \pi\left(s_t\right)$$

The value of a given action depends on the state in which it is performed, and the time it was taken. Reinforcement learning runs the agent through sequences of (s_t, a_t) pairs, noting the resulting rewards, and calculating the Q-function until it produces the best trajectories for the agent to take to maximize the *Goal* in various situations and through many games, in effect establishing the policy function π.

The policy function must of course avoid simply repeating the same actions or moves that previously garnered the highest rewards (overfitting), for that may cause the agent to forego actions with possible higher rewards, so in addition to *exploitation* of old avenues, *exploration* of new fields should be included in the algorithm; the ratio of the two is

$$\varepsilon = \frac{\text{Exploration}}{\text{Exploitation}}$$

where the more daring agents will have a higher ε value and be so-called *ε-greedy*.

For example, in the video game *Pong*, the paddle agent successfully hitting the ping-pong ball back has a positive reward of continuing to the possibility of gaining a point, and missing the ball has a negative reward of losing a point, with the goal of course being amassing a certain number of points before the opponent does.

Anyone who has played ping-pong, tennis, handball, or racquetball has encountered the doughty returner who just returns every shot and the attacker who takes chances with daring shots. The ε-greedy player's exploratory shots can result in immediate rewards of winning points but are easier to miss and result in a punishing loss of a point, whereas tried and true shots exploit the delayed reward of steady return play banking on an opponent's error. The reward function thus reflects the ε-greedy player's percentage of missed attacking shots.

After many games, the discount factor γ will have modulated the rewards commensurate with the player's skill level, as revealed by the iterative accumulation of reward and punishment over multiple games.

A computer vision convolutional neural network is employed to recognize a state, for example, the image of a barrier and its surroundings confronting Super Mario represents a state, and after the DCNN recognizes the barrier, then the policy function π based on those values maps a state s_t to the best-known action a_t. The ranks of the possible actions that the agent can perform in that state to overcome the barrier, for example, jumping over a barrier will give Super Mario 10 points because going around the barrier takes more time, it gives only 5 points, and hitting it head-on will result in a −5 points punishment.

It is important to realize that there are no physical principles at play in reinforcement learning and there is no need for supervised training (although it can be employed for a head-start). RL proceeds through Q-learning which chooses paths of actions iteratively that produce higher expected values based on the rewards and punishments of the action.

Another variation of reinforcement learning is the *policy* and *value* networks, which will be described in Chapter 17 on AlphaGo.

Reinforcement learning algorithms thus are supremely generalizable as they learn from the accumulation of own experience in a given situation just as humans do. Thus, like humans, they can explore and handle many different tasks completely bottom-up with no top-down hand-engineering, and since they have no subjective predilections, unlike humans, they can objectively proceed based entirely on the Q-learned optimal policy, and after tireless 24/7 practice, they can easily defeat the emotionally impaired and time-constrained human without even *a priori* knowing the rules of the game!

TORONTO/DEEPMIND VIDEO GAMER

The University of Toronto's Volodymyr Mnih and colleagues' goal was to create a single artificial neural network inference machine that is able to learn to play a variety of games.

The artificial intelligence *finite state* player matched against physical barriers and adversarial *non-playing characters* (NPCs) breakthrough was the Toronto/DeepMind Technologies' convolutional neural network reinforcement learning AI Video Gamer, which with no prior knowledge of the rules of the game, and from scratch defeated expert video game-playing humans in the highly competitive environment of the classic Atari games, *Beam Rider, Breakout, Enduro, Pong, Q*bert, Seaquest,* and *Space Invaders*.

The network was not provided with any game-specific information or hand-engineered guidance and was not privy to the workings of the Atari 2600 emulator that was used in the competition.

The Toronto/DeepMind Video Gamer learned from nothing but raw pixel video input, reinforcement learning, and it generated a set of possible actions using purely experiential replay memory. There were no adjustments of the architecture, learning algorithm, or hyperparameters for different games; the Toronto Video Gamer performed just like a human player across all seven games, clearly demonstrating a robust game-playing capability and achieved superior benchmark performance.

NOTES

1. MDP diagram by waldoalvarez, licensed under Wikimedia Commons by ttps://commons.wikimedia.org/wiki/File:Markov_Decision_Process.svg #file.
2. Program *nesting* places a computer command (for example, functions, DO-LOOPs, conditionals, etc.) inside another command; it is used in *Python* and *JavaScript*.

AlphaGo

THE EVENT THAT GALVANIZED the promise and prowess of artificial intelligence was AlphaGo's stunning victory over the reigning World Champion, Korea's Lee Sedol, in the ancient "surround and conquer" Chinese board game of *Weiqi* (known in the West by its Japanese name *Go*).

The 19 × 19 board and possible 361 stone placement positions are simply played by placing stones to surround and acquire territory while capturing your opponent's stones. Although in the first instance a seemingly simple game, the lower bound of 2×10^{170} possible combinations of positions (much greater than the 10^{120} combinations of chess), that is, indeed greater than the sum of all the atoms in the universe; in a reverse of Leonardo da Vinci's words, "simplicity is the ultimate sophistication". [1]

Faced with a truly astronomical number of static decisions and the genius-level creativity of *Go Masters*, mathematicians and computer scientists, many of them avid *Go* players, had always believed that a master-level *Go*-playing computer was an absolute impossibility.

Nevertheless, in 2010, a company in England, DeepMind succeeded in developing early *Go* computers using tree searches that could play at an amateur level, but in the intricate process of developing something as complex as a master-level *Go*-playing machine was fast depleting its funding budget. Fortunately, in 2014, the new tech giant Google came to the rescue and acquired DeepMind and together set out to challenge professional *Go Masters* from Europe, Japan, Korea, and China, just those paragons worshiped in East Asian societies as exalted members of the most sublime class of spatial analytical geniuses.

 DOI: 10.1201/9781003463542-17

In 2015, for the first time, Google's *AlphaGo* won a sanctioned match without a handicap against a professional, the European champion Fan Hui, who understandably impressed, immediately joined DeepMind as a consultant.

Confidence rapidly growing, AlphaGo next challenged and defeated the legendary Japanese champion Iyama Yuta (井山裕太), and in the very next year, challenged the reigning World Champion Lee Sedol (李世乭), the winner of 18 World Champion Matches and regarded as one of the greatest *Go Masters* of all time.

Born in a little town in one of the small islands adjacent South Korea's southeastern coast, few Koreans knew how to pronounce the third Chinese character in his name (李世乭), let alone what it meant. He gained professional status at the age of 12 and achieved *Go* Master ranking when only 19 years old. In 2016, he was at the height of his powers.

The match was held in Seoul at the Four Seasons Hotel ballroom in March 2016 and was streamed live to an estimated 200 million *Go* cognoscenti in South Korea, China, Japan, and America, plus mathematicians and computer scientists in England, Europe, America, and East Asia.[2]

Board-game computers were always thought to depend on deep searches and coldly logical processing in limited skirmishes ostensibly as might be discerned from part of the AlphaGo search tree process shown in Figure 17.1.

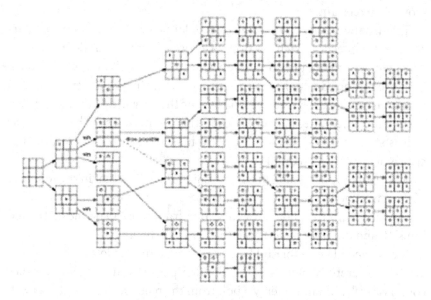

FIGURE 17.1 AlphaGo Monte Carlo Tree Search schematic.

They were thought to lack the creativity of daring play such as demonstrated by *Go Masters* like Lee Sedol, who although always aggressively looking for and exploiting local *fights*, where their clever manipulation of stones would disrupt, hem in, surround, and capture the opponent's stones in martial arts-like fights to win territory and capture stones.

But acutely aware of opportunities, a *Go Master* knows when to leave the *tesuji* (手筋) to foray into new areas to gain a hinterland initiative. Indeed, it was just the creativity of a *tenuki* (手抜き) that was believed to separate man from machine; that is, the mix of exploitation of the fight with daring exploratory ε-greedy forays as described in the Game Theory described in Chapter 16.[3]

Rather like Premier League football, the "beautiful game" is attractive because of the superb skills of individual players in instances of cleverly outfoxing defenders with deft footwork and precision passing, actions that are beautiful to watch but seldom contribute to a final victory, something that often is the result of the *value* of a winning strategy and of all the players' positional discipline during the match.

Unfortunately for the fans, this often results in stultifying defensive matches that thwart aggressive, well-organized, and crowd-pleasing shots on goal by the opponent, but wins through breakaway counterattack, in matches usually ending in a 0 – 0, or 1 – 0 scores. That is, the objective is to win, and at best tie a superior opponent; how many goals your team scored is irrelevant.

The usually modest and polite Lee Sedol before Game 1, out of character, guaranteed a victory over AlphaGo. Perhaps underestimating AlphaGo, he rapidly played his stones, seemingly without great concern. But when AlphaGo took an exploratory *tenuki* in the upper left of the board instead of the perceived wisdom of the "savoring" *aji* (味道) in a current local fight, Lee Sedol now knew that AlphaGo could be creative as well as strictly logical. It was a play that many commentators regarded as critical to victory in Game 1.

After Game 1, Lee Sedol, in explaining the loss, said that in the first part of the game, he felt that he had made some questionable plays, and while continually thinking about how to recoup from those plays, he distracted himself, and did not respond well to AlphaGo's *tenuki*.

A chastened Lee Sedol said he was surprised that he lost, and vowed to play more carefully. Both sides in the first part of Game 2 played rather conservatively, when suddenly a bolt from the blue astonished Lee, as well as the commentators who leapt to their feet in surprise.

AlphaGo's black 37 *fifth-line shoulder hit* seemingly gave up too much potential boundary territory to white, and indeed a post-game check of AlphaGo's play log found that the *katatsuke* (尖衝) was ranked at only a 10^{-4} value, yet even with Lee Sedol's excellent response, this exploratory one-sided *sente* on the opponent's diagonal was widely regarded as the key to victory because it unified AlphaGo's total board position by subsequent placements as shown in Figure 17.2.[4]

A Chinese maxim of *Go* play is "superiority lies in the center, inferiority is at the sides, and the commoner is at the corners" (高者在腹、下者在边、中者占角). In Game 2, AlphaGo demonstrated once again its superiority by tending to the center with black 37, and proved its creativity by daring to go against the perceived wisdom of a reckless play.

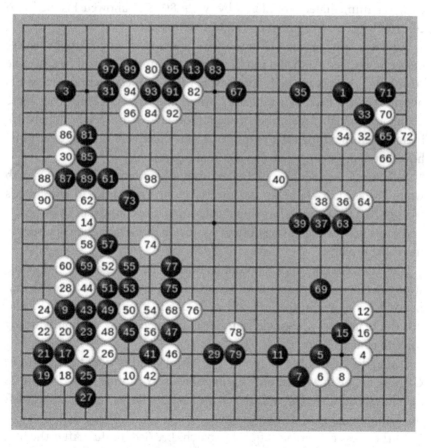

FIGURE 17.2 AlphaGo's "Shoulder Hit" vs. Lee Sedol Game 2.

The following Game 3 demonstrated AlphaGo's self-supervised reinforcement learning (RL) by seemingly correctly estimating the value of look-ahead play, and as in Game 2, AlphaGo's Monte Carlo Tree Search (MCTS) play expansion process was a demonstration of planning, similar to that of human masters of the game.

Lee Sedol's third straight loss was more than discouraging, as it apparently confirmed to him and his team of AlphaGo's superiority, but just as suddenly as AlphaGo's sublime shoulder hit *katatsuki* in Game 2, a human could also astonish the machine, and give humankind a spark of hope.

In Game 4, Lee Sedol's "divine wedge" white 78 *tesuji* in the center, although at first disparaged by AlphaGo's team, demonstrated his genius and courage. It can be surmised that AlphaGo's MCTS *selective node expansion* inadvertently skipped over the white 78 possibility as too remote and did not know how to respond, playing a desultory black 79 that was immediately countered by white 80, and allowed Lee Sedol to control the game from then on (Figure 17.3).[5]

AlphaGo lost perhaps because Lee's white 78, being so unexpected, discombobulated AlphaGo into a number of "mistakes", but because the "thinking" process of the hidden layers in an artificial neural network is not discernable even by its creators, so no one will ever know (but there is ongoing visual transformers research on CNNs that feed the network images that maximize the output of a particular neuron and determine how that neuron contributes to the final image).

In Game 5, Lee Sedol hoped to maintain the momentum of the superb white 78 *tesuji* of Game 4, and together with the AlphaGo entourage's worries that the Game 4 loss might cause it to lose confidence in its policy and value networks and succumb to confusion in the face of an opponent's improbable play.

But AlphaGo once more showed its superiority at the center and with a strong defensive posture, but rather ordinary play, with white 90 the key, Lee Sedol resigned in the face of a white 276 in a long, close Game 5 and lost the match 4 – 1. Google generously donated its $1 million winning prize to charity.

It may be that only a player of Lee Sedol's ability could persevere in Game 5, but he knew that only his extraordinary "divine wedge" white 78 in Game 4 had staved off total defeat, he ruefully smiled, and hurriedly left the Championship Match ballroom.[6]

In the press conference after the match, Lee Sedol said, "After the first game, I was surprised I lost, but from the very beginning of the second

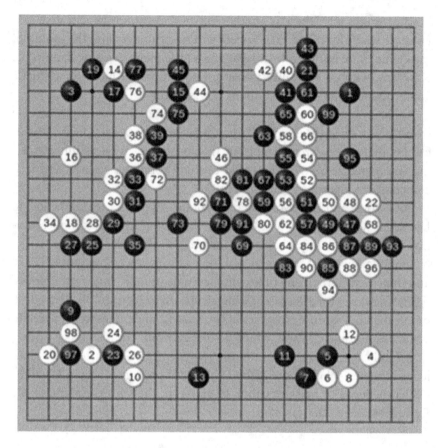

FIGURE 17.3 AlphaGo vs. Lee Sedol's "Divine Wedge" in Game 4.

game, I could never manage an upper hand for one single play. It was AlphaGo's total victory".

In a kind but futile attempt to sooth and reassure the 200 million humans all over the world who followed the match, he added that "it was my defeat, not a defeat for mankind".[7]

Although gracious, such words ring hollow then and now, for Lee Sedol, a child prodigy who gained professional rank at 12, won his first championship at 19, and was the No. 1 player in the world from 2002 to 2015, winning 18 straight championship matches; he was believed to be unbeatable. He was and is a national hero revered in Korea, and celebrated throughout the Go-playing world. He represented mankind nobly in a match with a machine, and he lost.

After earnestly taking full blame for defeat at the hands of a machine, a shaken Lee Sedol retired in 2019, plaintively realizing that no matter how hard he might try, "there is another entity that cannot be defeated". He however wistfully added that "robots will never understand the beauty of the Game as we humans do", displaying a little pique in losing, but unlike Kasparov, with an equanimity that only questioned AlphaGo's aesthetic appreciation of the game and not its playing skill.

AlphaGo's stunning victory, although roiling the waters of the *Go* world, also turned the limelight onto the hitherto backwater field of artificial intelligence research, completely thawing the AI Winters, and attracting young people all over the world to study computer science, and tech companies to vigorously recruit them.

But Google and DeepMind were not done, AlphaGo's sole game loss to Lee Sedol was attributed to a "delusion" resulting from program saturation ignoring Lee's improbable "divine wedge" white 78.

In the following year, armed with four tensor processing units (TPUs) and improved reinforcement learning capable of more innovative moves, *AlphaGo Master* (one branch of which is shown in Figure 17.4), won 60 – 0 online matches against professional *Go Masters*, and prizes were offered to anyone who could defeat it, which no one ever did.

An unbeatable and constantly improving AlphaGo Master then challenged the new World's No. 1, the 18-year-old prodigy Ke Jie in Wuzhen, China, believed to be the birthplace of *Weiqi*, the Chinese name of the Japanese-named "*Go*".

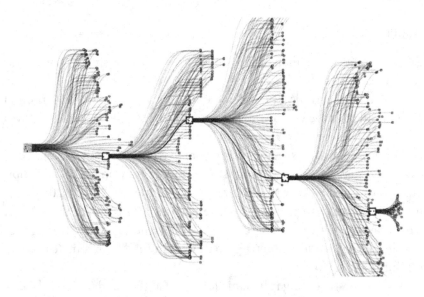

FIGURE 17.4 One Branch of AlphaGo Master's expansion.

Ke Jie at first had been reluctant to accept AlphaGo Master's challenge, not as he said from fear of losing to a machine, but rather because he was afraid it would "copy my style", and indeed, AlphaGo Master had undergone supervised training on *Go* Master match publications, and its match against China's young genius for its part, very likely was a chance for another high-level learning experience.

AlphaGo Master, however, did the teaching rather than the learning, and won 3 – 0 for a US$1.5 million prize. After the Match Ke Jie attributed his defeat to his inability to counter AlphaGo Master's "non-human playing style", adding a doleful prospection:

> *It seems that from now on, Go will be dominated by machines. After humanity spent thousands of years improving our play, a machine tells us that we were completely wrong; we apparently have only scratched the surface of the essence of Go, but machines will never have the human respect and love for the Game.*[8]

Europe, Japan, Korea, and China's professional *Go* associations subsequently all awarded the honorific highest 9-*dan* certification to AlphaGo, thus bestowing a *Go* Master's ultimate rank to a machine.

After the epochal match with AlphaGo, a chastened, but now more formidable Lee Sedol went on to win all of his subsequent matches against human *Go* Masters, stating that conversely to what Ke Jie had feared, Lee Sedol "learned from AlphaGo", and that "it has changed the way *Go* would be played in the future". That is, the machine had something new to teach even the supremely expert human in accord with the board game adage:

> *Sometimes you give them a lesson, sometimes they give you a lesson.*

Lee Sedol and Ke Jie thereby revealed a cultural difference between an accepting East and a recalcitrant West, represented by Kasparov and his fans, that portends smoother acceptance of artificial intelligence in Northeast Asia than in Europe and America.

Unsurprisingly, but like championship chess, if computers are the best players, won't human matches lose their appeal? Will *Go* Masters be replaced by computer scientists and their machines playing against each other for real *Go* supremacy?

ALPHAGO TECHNOLOGY

AlphaGo learned to play following the standard machine learning process, first from supervised training from a set of published professional *Go* matches, and then improved play by reinforcement and self-supervised learning from many simulations. A deep convolutional neural network (DCNN) is used to present board configurations, and an advanced Monte Carlo Search Tree (MCTS) algorithm with a *policy network* to determine the next play and a *value network* to predict the play's contribution to winning the game.

AlphaGo used an extension of a Markov Chain that allowed more choices and reinforcement rewards; this Markov Decision Process (MDP) employs the policy network where a policy function π is a probabilistic mapping from the state s of the *Go* board to an action in respect of that state $\pi(s)$. When an MDP is combined with the policy network, the action for each state is determined and the MDP is a regular Markov Chain with a Markov transition matrix.

AlphaGo's idea is to choose a policy π that will maximize the sum of the rewards $R_{\pi(s_t)}$, over time t multiplied by a discount factor $(0 \leq \gamma \leq 1)$ raised to the power t, where s_t is the state at a given time t and $\pi(s_t)$ is the action governed by the policy π (a_t in Chapter 15); and, after training and RL the *optimal policy* π^* is given just as in reinforcement learning by maximizing a summation over rewards and discount factors for progressing states of the game, where a lower γ motivates a relatively immediate action, and for the policy network, it is usually closer to *1*, reflecting the typical long look-ahead strategy in *Go*.

$$Goal = max \sum_{t=0}^{\infty} \gamma^t R_{\pi(s_t)}\left(s_t, s_{t+1}\right)$$

The Monte Carlo Tree Search used by AlphaGo requires a generative model that includes the stone distribution and how likely a play will affect it (similar to generative models used in NLP to determine next words).

The algorithm to calculate the optimal policies of MDPs uses the policy network π for actions and the value network V of real numbers; π will have the action solution and $V(s)$ will show the discounted sum (a real number) of the rewards that will be earned on average by computing that solution from state s.

The policy and value networks then will be updated and iterated until there is no change and is thereby recursively computing a new estimate of the optimal policy until convergence to a truly optimal policy π^* and the highest state values for a win.

Every MCTS node is a board configuration, with the *edges* of the node being the possible plays. The MCTS expands the search tree by random sampling of nodes and simulates the game through the Monte Carlo long sequences of alternate plays (*simulation of sample paths*) to the end.

Starting from the root of the search tree, children are selected until a leaf node (no children) is reached, and then by *selective node expansion*, MCTS creates more child nodes. Then the Monte Carlo *playouts* (also called *rollouts*) are backpropagated to update the *value*.

After many, many Monte Carlo simulation playouts, the final result of each simulated game is used to weight the nodes in the Monte Carlo Tree Sarch *according to the number of wins* produced using those nodes. From 3 million reinforcement learning games, AlphaGo established its optimal policy π^*.

The self-supervised reinforcement learning refines the value network to select the children node paths that lead to the most victories, and these nodes are used in the future simulations, thus improving play with each simulation.

Values are based on the study of *Operations Research* estimating *Upper Confidence Bounds* (UCB, *UCTs* for search trees) that are sequential decision-making algorithms. These evaluations can be based on immediate exploitation because of high average win rates in many immediately preceding simulations, or exploration of new plays after only a few good simulation results.

Thus, a given board configuration estimation of the play *policy* and the *value* of winning the game are determined by the parameterized value network that makes evaluations based on approximating the value of a play and a probability distribution for the opponent's play, the probabilities calculations typically starting with a Bernoulli distribution with given rewards,[9]

$$UCT_i = \frac{w_i}{n_i} + c\sqrt{\frac{lnN_i}{n_i}}$$

where w_i is the number of wins for the node after the ith play, n_i is the number of simulations for the node after the ith play, N_i is the total number of simulations run after the parent node (considered after the ith play), and c is the exploration parameter, theoretically equal to $\sqrt{2}$, but often chosen for the particular situation.

The first term measured exploitation, being high for plays with high average win ratios; the second term measures exploration, high after only a few simulation runs.

If the first term in the equation above w_i/n_i is high, the large number of local fight wins to that point is also high, and AlphaGo should continue exploitation of that local fight; conversely, if w_i/n_i is low, a smaller number of wins means that AlphaGo should leave the local fight and explore other areas of the board. If the second term is high, the simulation search likely has *expanded*, and AlphaGo has more opportunities in a broader search.

Go experts generally see machine play as coldly logical and thus prone to the close-quarter analysis of the intricate local fights of *tesuji*, and seldom venture *tenuki* plays, but the UCT_i considers *tenuki* for AlphaGo. Top players like Lee Sedol and Ke Jie generally aggressively seek local fights to capitalize on their adroit *tesuji* play, but unlike the ordinary *Go* player, they can better recognize promising *tenuki* opportunities and will opportunistically leave the local fight.

AlphaGo's data structure policy network comprised 12 hidden layers of convolution filters with *zero* padding and stride 1 to maintain the spatial dimension. The network takes $19 \times 19 \times 48$ input features to represent the 19×19 board. The input layer uses $5 \times 5 \times 49 \times 192$ filters while the hidden layers all use $3 \times 3 \times 192 \times 192$ filters, and all the layers are rectified by a sigmoid or other greyscale function. The last convolutional layer is a $1 \times 1 \times 192 \times 1$ filter with different biases for each location followed by a *softmax* function.

The value network is similarly constructed, but hidden layer 12 is an additional convolution layer, layer 13 is a $1 \times 1 \times 192 \times 1$ filter and layer 14 is a fully connected layer with 256 *softmax* rectifiers. The output layer is a fully connected layer with a single *tanh* output.

AlphaGo's policy network went through 30 million different board configurations, needing 50 GPUs for three weeks of training. The value network used 50 GPUs and one week of reinforcement learning.

The MCTS simulations initially used 48 in-line CPUs, but later changed to massively parallel GPUs. AlphaGo's hardware comprised 1,920 CPUs, and 280 GPUs, and in the matches against Lee Sedol and Ke Jie, employed Google's process-accelerating tensor processing unit (TPU) as ASICs.

The strategy is like the football manager's single objective of winning the game and choosing the best winning *value* whether it is entertaining or not. The reinforcement learning uses the γ discount factor to select optimal plays for a given situation.

For AlphaGo it is not important *how* you win or *by how much* you win, or even *how you play* the game, the only goal is to win. If AlphaGo must choose between a scenario where it will win by six goals with 80% probability and another where it will win by one goal with 99% probability, it will choose the latter.[10]

AlphaGo's algorithm is complex and subject to scrutiny, but is based on mathematics and executed by computer programs in optimization simulations. Whether it all works is confirmed by, among others, Lee Sedol and Ke Jie.

ALPHAGO MASTER, ALPHAGOZERO, AND ALPHAZERO

After AlphaGo Master's 60 – 0 straight wins against other *Go* Masters and defeat of Ke Jie, might AlphaGo Master gain a confidence that could easily devolve to conceit?

It is of interest to note that some expert commentators have found that conversely to Ke Jie's description of its "less human style", AlphaGo Master might be *more human*; that is, AlphaGo Master developed a confidence that its choices were always the best because of its value network, otherwise it would not have made them. AlphaGo Master might consider a human opponent's play clever, not bad, but nevertheless inferior to a response born of a superior (toward winning) *value* riposte, and such successes could easily be interpreted as "intelligent arrogance".[11]

Since AlphaGo Master could defeat any *Go* Master, it might have felt that there was no need for a teacher, so AlphaGo Master stripped off what in comparison it considered a plebeian training set module, and approached Go as a pure-play *inference engine*.

The new, refreshed *AlphaGoZero* could learn how to play any game from just playing the game on its own; that is, starting from random play, from a combined more effective single policy and value network, and using a simpler tree search algorithm, by self-supervised reinforcement learning, and an advanced *polynomial UCT (PUCT)*, had no need for Monte Carlo rollouts, but could still employ them just to be sure.[12]

The new reinforcement learning algorithm incorporates look-ahead searches inside the RL training loop. In each state, a Monte Carlo Tree Search (MCTS) is guided by the combined networks, and produces the π probabilities of each play more robustly, and as such is a more powerful policy network creator. Self-supervised learning with tree search uses the MCTS-based policy to select each play, then using a sample value game winner produces a more powerful policy.

The tree searches are then iterated, and the network's parameters are updated to more closely match the improved search probabilities and determine the best value. The new parameters then are used in the next iteration of RL, thus making MCTS even stronger.

There were three days of 4.9 million self-supervised RL training games with 1,600 MCTS simulation runs. AlphaGoZero reached AlphaGo performance that took three months in only 36 hours. [13]

AlphaGoZero in competitions with supervised AlphaGo, although starting slower, still defeated AlphaGo in 24 hours, and produced new *Go* plays never before seen in professional play, implying a superhuman capability.

The zero in "AlphaGoZero" means that it starts from random input, and with 64 GPUs and 19 CPUs, runs on the *TensorFlow* platform, and by 100 – 0 defeated AlphaGo Master.

AlphaGoZero's self-supervised reinforcement learning could not just involve playing the same version of itself over and over again, because the algorithm would *overfit* the same version and much like a pedestrian player, merely memorize responsive plays rather than creating new strategies and tactics.

AlphaGoZero was invincible because different versions of itself were not only each in turn the best possible, but the almost infinite loop of iterative improvement meant that AlphaGoZero would develop its skill such that it could never be defeated and become *infinitely* good at *Go*, whatever that means and portends. It was unbeatable except by some other machine that somehow got ahead of AlphaGoZero's learning curve.[14]

Not limited to *Go*, AlphaGoZero's successor *AlphaZero* employing 5,000 first- generation TPU, 64 second-generation TPUs, and 44 CPUs, running MCTS RL simulations, played professionals, indeed did quickly learned to play not only *Go*, but also checkers, chess, *Xiangqi*, and *Shogi*, and beat human champions in each of those board games after only a few hours of self-supervised reinforcement learning.[15]

The AlphaGo series of *Go* algorithms, through relentless self-study reinforcement learning could determine optimum board position strategies as it iteratively increased the probabilities of ultimate victory against the best player in the world, itself. This classic instance of diligent self-strengthening leading to a *Go virtue* through comparison with oneself is quintessentially Confucian, and although developed in England and America, is entirely proper for a *Go* Master in Northeast Asia.

NOTES

1. The number of atoms in the Universe, composing the galaxies, stars, planets, and interstellar gases (mostly hydrogen) is estimated at 10^{78}–10^{82}, but in truth those atoms only make up about 4% of the total mass in the Universe, with the rest residing in the mysterious *dark matter* (16%), that cannot be seen and has not yet been identified but nevertheless holds galaxies together, and the mass-convertible *dark energy* driving the expansion of the Universe. See Chen, R.H., 2017, *Einstein's Relativity, the Special and General Theories with their Cosmology*, McGraw-Hill Education (Aisa).

2. Image from Philip Harvey Photography Inc. *Independent News*, public domain because diagrams consist of only known geometrical shapes or text and lacks copyrightable originality.

3. Because of the motivation behind certain plays is known only by the player, *Go* terms in Japanese, Chinese, and English are not precisely distinguishable and may be used differently.

4. As opposed to chess, after many reverses and much partiality among Japan, China, and Korea, black plays first in *Go*.

5. All board figures from Wikipedia interactive are free use because they consist of only known geometrical shapes or text and lacks copyrightable originality.

6. A beautifully done documentary film on the AlphaGo–Lee Sedol match is *AlphaGo Documentary* on YouTube, provided by Rajarshee Mitra.

7. Lee Sedol quote from Byford, S., 2016, *The Verge*. AlphaGo Master Search Tree branch image from Google, *Library of Congress*. Lee Sedol quote from Le Roux, M. and Mollard, P., 2016, phys.org.

8. Ke Jie's quotes *D. Silver et al.* 2017 (7676): 354.

9. A Bernoulli distribution gives the probability of achieving a "success" or "failure" from a Bernoulli trial with only two possible outcomes (success or failure).

10. Chouard, T., *Nature*, 12 March 2016.

11. Ref. Daniels, B., 2017, *Whatever You do Is Wrong*, YouTube.

12. PUCT is a consistent sequential decision-making algorithm that does not require knowledge of the action space; the only assumption is a transition matrix and a black box action sample; Rf. Auger, *et al.*, HAL Open Science, Sept. 2013, "Continuous Upper Confidence Trees with Polynomial Exploration."

13. Silver, D., *et al.*, *Nature, vol. 550*, 19 October 2017.

14. AlphaGoZero is like the genius Einstein, who after recognizing his superiority, skipped classes and through thought experiments presented problems to himself which he solved by himself and thereby created the whole new fields of relativity physics and cosmology. The innately talented Michael Jordan also relentlessly practiced to improve to levels of basketball skill even when he had no equal.

15. For details of the algorithm Silver, D., 2018. AlphaGZero's learning curve for each game is shown in this video.

CHAPTER 18

Game Theory

G ENERAL GAME THEORY USES mathematical models of strategic interactions between rational agents. Essentially stated, John von Neumann used L.E.J. Brouwer's fixed-point theorem, which states that continuous mapping into a space that has no *punctures* or missing endpoints (*compact space*), or a subset that intersects every line into a single line segment (*convex* set) to prove that there is an equilibrium in a two-agent zero-sum game, where each agent's gains or losses are exactly matched by the losses and gains of the other agent.

For example, rational numbers have *punctures* (the holes are the irrational numbers such as π and e) between numbers; real numbers are an open set (x, y) and the set is not *compact*, whereas the closed interval $[x, y]$ set is *compact* because it has the limiting numbers. Given any two points in a *convex subset*, the subset contains the whole line that joins the two points, for example, a solid cube is a convex subset and an empty cube is not.[1]

Seemingly abstruse, if one thinks about it, it does form a basis for a two-player, zero-sum equilibrium in games. For example, a fair card game has no punctures (like a loaded deck or cheating), so it is *compact*, and the alternating plays are points on a convex subset that have some relation completely within the rules of the game.

John Nash extended von Neumann's *strategic equilibrium* to no restrictions on payoffs or the number of players, and it was applied to economics and games (the *Nash equilibrium*), and he proved that every finite game has a strategic equilibrium. In a finite zero-sum game, a Nash equilibrium

---

DOI: 10.1201/9781003463542-18

	B		
	ROCK	PAPER	SCISSORS
A ROCK	0,0	-1,1	1,-1
PAPER	1,-1	0,0	-1,1
SCISSORS	-1,1	1,-1	0,0

FIGURE 18.1   Rock–paper–scissors strategic form.

is reached when every player employs a strategy such that no one player can benefit by changing strategies.

Although there is no mathematical theorem for always winning, if the Nash equilibrium can be discerned by either player by observing playing behavior and seeing the results, then at least there will be a tie if not a win.

For example, in the simple two-player game of tic-tac-toe, if player A puts his "X" on any corner, and B just randomly plays his "O" except in the center, the best B can do is draw, and he will lose if A adheres to the Nash equilibrium. Once the players realize that a center "O" in the first B move will always end in a draw (if the players play rationally), it clearly demonstrates that there is a strategic equilibrium, and the players can play an infinite number of games and always end up in a draw.

Another simple zero-sum game, but without perfect information, is *rock–paper–scissors*. The payouts where winning is *1*, losing is *−1* and a draw is *0*, can be displayed on a *strategic form* shown in Figure 18.1. In this finite, zero-sum game, the Nash equilibrium is reached when every player employs a strategy such that no one player can benefit by changing strategies.

It turns out statistically that if one player randomly plays each choice, 33% of the time, in the very long run, that strategy cannot be exploited, and if the opponent discovers that and plays the same way, neither player can exploit the other, and over many games, the players will have about the same number of wins and losses.

But if one player diverges from the equilibrium strategy, for example, increasing the percentage of scissors plays, although winning in some instances, that player's departure from the von Neumann equilibrium can be exploited, and in the long run since he cannot exploit an opponent who strictly adheres to the equilibrium strategy, he will lose more than he wins because he withdrew from the strategic equilibrium.

TEXAS HOLD'EM POKER

Aside from the recondite mathematics of Brouwer and von Neumann's game theory, and the professional matches of chess and *Go* held in somber chambers of refined intellectual competition, high-stakes poker, although based on mathematical probabilities as well, is played mostly by gamblers in raucous casinos.

So far the machine has bested humans in contests of logic, speed, and memory, but those are what humans already regard as the computer's strength; can a machine beat man in the domain of imperfect information, for instance the quintessentially human game of bravado, cunning, baiting, deceit, bluff, intimidation, and subterfuge that is *Texas Hold'em* poker? A poker game victory would demonstrate that a machine can be vulgar as well as refined.

In 2017, a perfectly poker-faced computer from Carnegie-Mellon University soundly defeated four top-ten professional players each in 120,000 hands of *heads-up* (two-players, so that there is no possibility of collusive ganging-up on a player as can happen in multiple-player games), *no-limit* (can bet total of chips owned at one time, making for much greater betting variability and risk) Texas Hold'em, with Professor Tuomas Sandholm and graduate student Noam Brown's *Libratus* ahead by $1,700,000 in simulated chips at the end of the 20-day competition.

Libratus employed game theory mathematical models of strategic interactions between rational decision-makers, something seemingly at odds with the daredevil machismo of poker, but in truth directed at the calm assessment of the winning bet probabilities necessary for victory.

Indeed, behind the façade of bravado lies a cold-faced logic that parlays the luck of the draw with the skill of the bet using *Nash equilibrium game theory* and *Monte Carlo simulations* to determine probabilities.

But with respect to artificial intelligence decision-making, a more essential difference is that in board games, the players know the exact position of all the pieces at every point in the game in a *perfect information* competition; whereas in card games, the players neither know the opponent's cards nor the order of card-dealing; the card games progress in an *imperfect information* game setting.

It is generally felt that a computer needs reams of informational data to process to a valid result, if the information is incomplete, can the computer processing give wunning results?

In an adversarial two-player, zero-sum game, even in the bravado yet insidious game of heads-up, no-limit Texas Hold 'em poker, the best players do behave rationally and thus satisfy the basic conditions of the Nash equilibrium.

In-game, the players can pass (unless another player has already bet), match, raise, or fold (collectively termed *bet*). The dealer first deals two face-down *hole* cards to each player, then lays out three *community cards* face-up, and the players bet; then a fourth community card is laid out, the players bet again, and after a last community card is laid out, a last bet is made; that is, the players bet depending on the combination of their two *hole* cards and three of the five *community cards*.

A computer cannot display feigned anguish or joy to deceive an opponent when dealt a hand or during betting, but by the same token, it cannot sense the opponent's facial and body language subterfuges either (although controversial advances in convolutional network facial recognition of emotion may soon make this possible).

For now, both the computer and its human opponent must rely solely on the *bets* to conceal one's own hand, and discern the opponent's hand.

Is there a scientific basis for developing a strategy that can be performed by a computer? Winning is of course based on probabilities in a constrained system that can be gauged and exploited, something that every poker player knows all too well, and a computer can calculate probabilities very well.

Betting strategy can be based on the Bayes formula and probabilities assigned to Markov Chain nodes in Monte Carlo tree-search simulations. Expert card players will always rely on the probabilities, such that they can ascertain, but since there are 10^{161} decision points in a game of poker, even an AI machine cannot traverse an entire game tree even once, and a deterministic choice for each play is impossible, and so the Nash equilibrium *for the whole game itself* is almost impossible.

Therefore, a model of the *abstraction* of cards and action is developed, which in computer science jargon means the removal of physical, spatial, and temporal details and attributes in order to focus on the essentials of the task at hand, in this case the bluffs and betting in a game of poker.

So, both the computer and its human opponent must rely solely on the raise amount, betting decisions and bluffs, and observe the opponent's betting to detect patterns. A player obviously must mix up his betting strategy throughout a game to avoid patterned behavior (overfitting) being recognized and exploited by the opponent, and likewise be able to recognize any patterns of betting behavior of the opponent, and adjust his own betting strategy (generalizable fitting).

The finite, adversarial, imperfect information game of poker lies in the exploiting the *Nash equilibrium* not from the game itself, but from that *abstraction* of the game, where there are many strategically similar situations that can be classified together for tree search, for example, similar hands like early-round king-high and queen-high flushes (*card abstraction*) and similar bets like $500 and $550 can be classified together in increments of $100 (*action abstraction*).

Libratus' first of three main modules computed approximate Nash equilibrium solutions using minimax algorithms in the poker abstraction to serve as a game *blueprint* for the early rounds of the game which in turn served as a precursor strategy for later rounds. The initial betting actions in the abstraction were patterned after the most common bet actions by the top contenders in the *Annual Computer Poker Competition* (ACPC), to provide the *strategic form* matrix elements. If during play, the opponent chooses an action that is not in the abstraction, that action is mapped to a similar action that is in the abstraction.

The blueprint was honed by Libratus playing simulated games against itself in reinforced learning using *Monte Carlo Counterfactual Regret Minimization* (MCCFR), an iterative algorithm that independently minimizes "regret" at every decision point, registering how much regret there is at not choosing an action in the past (anyone who has ever played poker knows the value of a MCCFR). So given the opportunity, Libratus will choose the action with the highest regret, then gradient-descent, backpropagate, and after many game iterations, the average of the regrets is minimized to approach zero, thereby improving the blueprint strategy.

In simulated games, one player will explore every possible action in the abstraction and update his regrets while the opponent plays solely on his current regrets. The roles of the two players are then reversed after each hand. An objective probability distribution is thus determined by the regrets of actions in previous games, thereby providing the *value* of a betting action. That value also depends on the probability of it being played in a later hand, where the value decreases if overly or underly played.

In heads-up but *limited* bets poker-playing systems, there are only 10^{13} unique decision points; if both players employ the MCCFR, their average strategies converge to a Nash equilibrium. But to account for the 10^{161} static decision points in *no-limit* Hold'em poker, Libratus improved the MCCFR by a sampled form of *Regret-Based Pruning* (RBP) where the high regret branches are pruned first.

The whole game then is broken down into *subgames* that are individually amenable to Nash equilibrium calculations which are used primarily for *defensive* purposes in finding one's own weaknesses to avoid being exploited by the opponent, and secondarily for *offensive* purposes of finding in the opponent's weaknesses and exploiting them. The strategy blueprint then can be iteratively updated.[2]

In game theory, a *subgame* is any subset of a game where all members of the subset belong to the subgame that has a single initial node and includes all of its own successor nodes as shown schematically in Figure 18.2 where there are altogether six subgames, two of which contain two subgames each enclosed by the ovals.

A subgame therefore is a game within a game, and by dint of its designed isolation, credible threats germane only to the whole game are in principle eliminated in the subgame, thereby allowing the player to concentrate on analyzing one part of the game.

In a two-player sequential game, player A chooses an action to go up or down at the whole game initial node. Player B then can choose to go left or right in a *subgame* within the ellipses depending on the action chosen by Player A. Then a strategic form matrix may be constructed of probable outcomes of the subgames, and a Nash equilibrium can be calculated using minimax for each subgame.

In Libratus' second module, a *nested safe-subgame solver* provides strategy in the subgame from an estimate of the value of reaching the subgame Nash equilibrium. The playing strategy blueprint module already estimated its Nash equilibrium and thus this value, and also for every subgame using these values for input, so the subgame solver solves in a finer-grained abstraction that is reached in real time; that is, it solves a new subgame every time an opponent chooses an action that is not

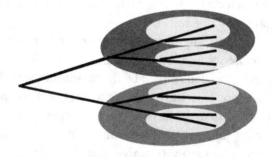

FIGURE 18.2   Single initial node and successor sub-games.

in the fine-grained abstraction, effectively including that action every time the opponent bets, thereby automatically and repeatedly calculating more and more finely grained detailed blueprint strategies as the play progresses.[3]

In actual playing however, a subgame should not be solved in complete isolation because winning strategies may depend on prior subgames and games hitherto not yet reached. If the blueprint is slavishly followed in this so-called *unsafe-subgame solving*, an opponent can recognize the patterns as simple isolated gambits and exploit them considering whole game experiences.

To offset this, a *safe-subgame solving* still places all actions within the blueprint, but a more detailed subgame abstraction using minimax aims to make the opponent worse-off, no matter what cards are held, by approximating an optimal strategy through assessing how much more a player would lose against a worst-case action by an opponent than if he simply followed the strategy blueprint, and in this way can considering bluffing and deceptive betting.

Libratus employs a dense action abstraction in the first two betting rounds of a game; in the *self-improver* third module, the missing branches in the original blueprint are filled in and a game-theoretic strategy is computed for those branches using the opponent's actual actions to guide the tree-search filling-in node values. If the opponent does not bet an amount that is in the abstraction, the bet is rounded off to a nearby size that is in the abstraction; this however causes a slight distortion in the strategy and estimates of reaching certain subgames, and the rounding error must be reduced by adding a small number of actions to the abstraction.

Which actions are added depends on the most frequent actions chosen by the opponent and how far those actions were from the solution in the abstraction, thereby filling in effective missing branches in the blueprint. Once an action is selected, a strategy with the new branches is calculated by the nested safe-subgame solver.

In this way, Libratus augments and refines the blueprint over time based on the *weaknesses in its own game* that the opponent has found in the strategy blueprint as determined by the opponent's actual play.

Libratus thus is not only learning how to exploit the opponent's play, but also learning how to make its own play less exploitable.

In an imperfect information benchmark AI challenge, Carnegie-Mellon's Libratus successively trounced four professional players in a heads-up, no-limit Texas Hold'em poker competition.

After two years of further development, Carnegie-Mellon's new *Pluribus*, not limited to two players, took on six players simultaneously in 15 no-limit Texas Hold'em matches and convincingly won them all. One would think that with the many more possibilities in multiplayer poker, Pluribus would need more computing power than Libratus' 100 CPUs provided, but Pluribus needed only two CPUs to defeat multiple top professional poker players.

The reason is that Pluribus took some pages out of AlphaGoZero's playbook on self-supervision reinforcement learning over *billions* of poker hands, experience that the best human players could never assemble in their entire lives. By starting from zero, just randomly betting, reinforced learning and then refining its play based on checking back after each training hand against itself as to which betting actions actually won the most money (like the value network in reinforcement learning), Pluribus thus used a *value* based on winning the game to decide its betting.[4]

In 2019 Noam Brown's new multiplayer *Pluribus*, in a six-player 12-day session of 10,000 hands of no-limit Texas Hold'em, defeated 15 top professional players.

The multiplayer game is more reflective of real-life situations of imperfect information, as in economics, business, stock markets, diplomacy, geopolitics, fraud detection, even warfare. The stakes are much higher, but given past human folly in those endeavors, in light of the demonstrated capabilities of *Libratus* and *Pluribus*, momentous decisions perhaps should be left to machines rather than humans.

After his victory, Professor Sandholm was asked by a reporter what game might be beyond a computer's capability, to which he replied that,

> *The imperfect information domain was the last*
> *Artificial Intelligence benchmark*
> *Our poker-playing machine has defeated the best human players*
> *and achieved superhuman performance.*

NOTES

1. Ref. Nash, C., and S. Sen 1991, *Topology and Geometry in Physics*, Dover.
2. There are Nash equilibrium charts and calculators for poker players' use, refer to poker.stackexchange.com. For the intriguing story of John Nash, refer to Nassar, S., 1998, *A Beautiful Mind*, Simon & Shuster, also a movie directed by Ron Howard and starring Russell Crowe (2001).

3. "Nesting" in computer programming means having an operation (such as a do-loop or conditional) inside another operation. Details or heads-up, no-limit Texas Hold'em algorithm, examples of play, proofs of theorems, and the computer program for the safe nested subgame solver, see Brown, N., and T. Sandholm 2017, *Science Research Articles*, doi:10.1126/Science. aao1733.

4. N. Brown and T. Sandholm 2019, Science, 365, 6456, 885, 30 August.

CHAPTER 19

Predictive Analytics

ARTIFICIAL INTELLIGENCE PREDICTIVE ANALYTICS deduces correlations in data to predict future performance from past behavior, there is no causation without correlation, but of course correlation does not *imply* causation, it might be just coincidence, and to the technologist, *coincidence* is just *noise*.

For example, if a veteran NBA player with a lifetime 3-point shooting average of 35% has made all of his six 3-point shots in the first half of a game, the opposing coach will tell his shell-shocked team, "Don't worry, he can't keep that up!"

If the coach means that because the shooter has made six straight shots, he more likely will miss the next shot, he is engaging in the *gambler's fallacy*, for any given 3-point shot is *stochastic*, meaning that a shot success does not depend on the success or failure of the previous shot.

If the coach means that the shooter will cool off in the second half, he is only slightly more accurate as the shooter is on his way to regressing to the mean, but if he means the shooter will absolutely *regress to the mean* during the second half, his sample size is too small to make any predictions. If the coach means that the shooter will regress to his average over the whole NBA season or for the rest of his career, he is correct in his assessment by *the Law of Large Numbers* (but of scant solace to his team in the present game).

The shooter may remain *on fire* throughout the game, but in games to come, he will cool off with a confidence interval as measured by the standard deviation in the 90% range; that is, the veteran 35% shooter will never perform in the larger sample size of whole season at greater than

DOI: 10.1201/9781003463542-19

a standard deviation from his average, meaning that no player, however great, can escape the regressive clutches of *Bernoulli's law*, and he will sooner or later experience a slump of say missing 10 attempts in a row.

But if the shooter increases his 3-point shot in practices for an hour every day, and his percentage rises to 43%, there may be a correlation or even a causation, and if the team's winning percentage also rises, there may be another dependent variable causation for more wins, which requires multivariate analysis.

More practically, predictive analytics is useful for predicting hardware and vehicle breakdowns in manufacturing and transportation, personal health problems, epidemics, and any correlation with a possible causal relationship.

In this sense, the real value of predictive analytics lies not in simple correlations such as the end of summer and the increased sales of school supplies, which are readily apparent and demonstrated by years of retail statistics, but in discovering *latent* factors and inferences from the statistical data.

The simplest regression model is to fit a straight line that best bisects the data points as shown in Figure 19.1 and a high variance data points distribution that a straight line would not accurately model the data as shown in Figure 19.2.

A data set with one independent variable x, and one dependent variable y, in a linear regression model is defined by the equation of a straight line,

$$y = mx + b$$

where m the slope of the straight line, and b the y-axis intercept.

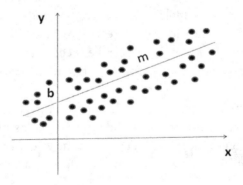

FIGURE 19.1   Straight linear regression.

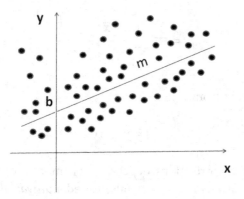

FIGURE 19.2   High variance data.

Once the slope and y-intercept are determined from the data, from any input of a new independent variable *x*, the *y* output can be predicted, not necessarily completely accurately for that single value of *x*, but a value that will equal an average over many *x* inputs; that is, a large sample size.

A linear regression model can also help to quantify the strength of the relationship between the independent and dependent variable by the variance of data points from the straight line; that is, if almost all the points are close to the line as shown in Figure 19.1, a strong dependence is indicated, and if the points are randomly scattered relatively far away from the line, a weak dependence is indicated.

If the distribution of data points is more diverse, such as in Figure 19.2, a different function such as sine or cosine or polynomial function might fit the data better.

Linear regression uses the *least squares* method, which minimizes the *variance* which is the sum of squares of the errors, to find the slope and y-intercept. This is done by first calculating the mean of the *x* values and the mean of the *y* values from the data points,

$$\bar{x} = \frac{\sum_{i=1}^{n} x_i}{n}$$

$$\bar{y} = \frac{\sum_{i=1}^{n} y_i}{n}$$

finding the slope *m* from,

$$m = \frac{\sum_{i=1}^{n}(x_i - \bar{x})(y_i - \bar{y})}{\sum_{i=1}^{n}(x_i - \bar{x})^2}$$

and the y-intercept from,

$$b = \bar{y} - m\bar{x}$$

If there are more correlations depending on more independent (input) variables, *multiple linear regression* (also called *multivariable linear regression*), produces for a particular dependent variable y (output), a linear combination of the multiple input variables x_i each of which is factored by a *regression coefficient* β_i that weights the significance of that variable, and a *residual error* term ε which distribution function is used to adjust the values of the total y_i, if needed or desirable,

$$y = \beta_0 + \beta_1 x_1 + \beta_2 x_2 + ... + \beta_i x_i + ... + \beta_k x_k + \varepsilon$$

For more complicated correlations, higher powers of the independent variables x_i, such as x^2 and x^3, or the interaction of the independent variables, such as $x_1 x_2$, or even elementary functions of the independent variables, such as $sin(x)$ and $cos(x)$, can also be used to fit the data. However, fitting regression curves too closely to the data points can lead to overfitting and a loss of a predictive model's generalization.[1]

If there are n observations on the $k + 1$ independent variables, then for the ith dependent variable,

$$y_i = \beta_0 + \beta_1 x_{i1} + \beta_2 x_{i2} + ... + \beta_i x_{ij} + ... + \beta_k x_{ik} + \varepsilon_i$$

This can be written in compact vector and matrix form as a *general linear regression model*,

$$y = X\beta + \varepsilon$$

where y is a column vector holding the y_i elements, X is a matrix arraying the independent variables x_{ij}, operating on column vector β that holds the regression coefficients β_i, and ε is a column vector holding the error terms ε_i.

For example, a company wants to optimize the time expended in the delivery of soft drinks to vending machines; the independent variables are

(1) how many bottles to stock (x_1) and (2) distance driven by the deliverer (x_2) in each run to the machines. Employing statistics software packages such as SPSS, a plot command to produce a matrix of scatter plots can assess if there are linear relationships among the data. If there are, then the regression coefficients β_i and error ε_i can be calculated by Gaussian elimination (typically by means of the determinant of the matrices) using canned systems of simultaneous equations solver software. The linear relation of delivery time as a function of number of bottles and driving distance then can be found in a fitted regression model equation with calculated regression coefficients such as,

$$y_{est} = 2.341 + 1.616x_1 + 0.0144x_2$$

where y_{est} is the estimated delivery time, and it can be seen that the delivery time dependence on the number of bottles x_1 is much greater than the driving distance x_2.[2]

After presumably optimizing the process by reducing the number of bottles stocked per machine and driving to more machines in a delivery run, if after many runs there is no significant improvement in the company's overall delivery time, then there may be hidden correlations at work, such as the refill percentage requirement of different vending machines. This factor then can be included in the analysis to see if there is any decrease in the overall delivery time, and if so, this correlation could be included in the model as a new independent variable.

The general linear model can be expanded to handle the effects of interdependent multiple correlations in a *multivariate linear regression* model,

$$Y = XB + E$$

where the dependent variable Y is a matrix with each column having a row of estimations of each of the dependent variables y as functions of the weighted independent variables, and the independent variable X is a matrix with each column being a set of observations on one of the independent variables x which is a function of the other independent variables. B is a matrix of parameters to be adjusted for fitting the data, and E is an error (noise) matrix that is assumed to be uncorrelated across observations and follows a *multivariate normal distribution*.

For example, a medical researcher collects data on the measurable independent health variables: weight, blood pressure, and cholesterol level of a cohort population, and further data on red meat, fish, milk, and alcohol consumed by the cohort per week. A multivariate linear regression model can determine the interrelated correlations among each of the health and diet-independent variables, and the errors (such as false reports of low alcohol consumption) are assumed to follow a normal distribution.

Simple linear, multiple linear, and multivariate linear regressions can be distinguished by the scalar, vector, and matrix representations as shown in the equations above, and the coefficients of the terms will reveal the interrelationships.

Further parameter calculations, hypothesis testing, analysis of the models, and the extraction of further information from intermediate steps can become quite complex, and are the subjects of ongoing research.

Linear regression is widely used in science and engineering data analysis, and in the biological, medical, behavioral, economic, and social sciences in predictive analytics. Common applications are *trend estimation* where the derivative of the curve trajectory can represent a trend, for example, in epidemiology, the change in trend is revealed by the second derivative of a number of cases versus the time curve. If the second derivative decreases, although the first derivative slope is still positive (meaning the number of cases is still increasing), the decreasing second derivative flattens the curve and a peak can be predicted, predicting the peak and subsequent decline of infections.

In public health studies, a linear regression model found a direct negative correlation between a cigarette smoking independent variable and a smoker's lifespan dependent variable. In finance, the *capital asset pricing model* uses linear regression and *beta* (whether the stock is more or less volatile than the market as a whole) to quantify the systematic risk of an investment. In economics, linear regression is widely used in almost all areas of prediction from economic downturns to inflation, and in artificial intelligence, linear regression is one of the fundamental learning algorithms used in supervised machine learning.

NOTES

1. Even with higher powers and elementary functions of the independent variables x_i there is still a *linear* dependence on the regression coefficients β_i so the dependence is still called a "linear" regression.
2. Bremer, M., *Multiple Linear Regression*, Math 261A, mezeylab.cb.bscb.cornell.edu.

CHAPTER 20

Support Vector Machines

C LASSICAL STATISTICAL ANALYSIS SUCH as regression analysis relies on the Law of Large Numbers, which means that as the number of observations tends to infinity, the empirical probability distribution function inevitably converges to the ground truth distribution function. The recent successes of artificial neural networks have largely rested on today's unprecedented Big Data and Large Language Models (LLM) that provide the large number of numbers as training sets for machines and information from the Internet and the Cloud to learn to classify, predict, and process with ever-greater accuracy.

There are, however, still many disciplines and industries that just do not have sufficient data to meet the requirement of the Law of Large Numbers, so AI machines were devised to provide classification for those cases with limited data.

As early as 1963, the Soviet Union's Vladimir Vapnik conceived a *support vector machine* (SVM) that could identify, classify, and perform linear regression, even with limited data. At first, he used classification lines and planes to graphically classify, but after 30 years, and with the advent of the computer, in 1992, he expanded his model so it could classify in greater than three dimensions; that is, into Hilbert *vector hyperspace*, and in 2000, SVMs could learn from self-supervision.

At first the SVM processed data in the same fashion as artificial neural networks, but differently from ANNs, operated in *vector space* with data

DOI: 10.1201/9781003463542-20

grouped in *data vectors,* and from the *inner product* of the vectors, the scalar distance between data vectors can be computed, thereby classifying the data in Hilbert vector space. This unique property of vectors is often used in artificial intelligence modeling.

In two-dimensional vector space, the data vectors are classified by producing a demarcation line or curve, based on the scalar distance between vectors, that best divides the data vectors onto opposite sides of the demarcation.

As shown in Figure 20.1, on one side of the optimal line are data vectors represented by black triangles (▲). The data vectors closest to the line, because they are data vectors that are classified differently are on the other side of the line, and represented by black circles (●). The data vectors, of each kind that are closest to the different class line are the most "supportive" of the classification, and thus are called *support vectors,* represented respectively by the empty triangles Δ and the empty circles (○) in Figure 20.1. The *optimal line* is the line that separates the different classes of data vectors by the widest *margin,* designated by the dashed line in Figure 20.1.

The margin is therefore like the width of a street, and the wider the street, the more distinct the classification of data vectors. The width of the street is measured by a line from the support vectors orthogonal line to the optimal line.

If, however, the data vectors are scattered such that a straight line cannot clearly separate them, the data vectors can be mapped by a process

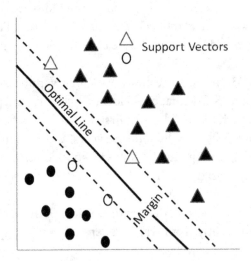

FIGURE 20.1 Straight line classification.

called *kernelling* onto a three-dimensional vector space by a "floating" in 3D space *flat plane*, and rotating the plane in space so as to best separate the circle and triangular data vectors to find an *optimal hyperplane* with the maximum margin of the support vectors as shown in Figure 20.2. The expansion to three-dimensional space has thus created one more degree of freedom for the separation of the data vectors into different classes.[1]

This kernelling of the space to four, five, six, and higher dimensions, and in principle even to an infinite-dimension Hilbert space, can be continued until a classification separation is manifested. But there may be many such hyperplanes and it would be best to find the *optimal hyperplane*, the one with the widest margins.

To find the hyperplane, first take the simplest two-dimensional case where the equation of the hyperplane is a linear function

$$y = ax + b$$

Now let $x = x_1$ and $y = x_2$, so

$$a x_1 - x_2 + b = 0$$

If we define vectors $x = (x_1, x_2)$ and $w = (a, -1)$, then

$$w \bullet x + b = 0$$

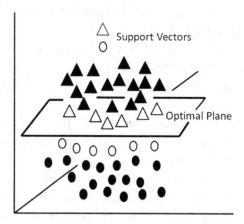

FIGURE 20.2   Flat plane classification.

Although derived from a two-dimensional case, because of the vectorization, this equation gives the equation of a hyperplane in any dimension.

For the *classifier*, define the hypothesis function h as

$$h(x_i) = \begin{cases} +1, \text{ if } w \bullet x + b \geq 0 \\ -1, \text{ if } w \bullet x + b < 0 \end{cases}$$

meaning that a vector data point on or above the hyperplane will be classified as class "+ 1", and a vector data point below the hyperplane will be classified as class "– 1". However, there may be many hyperplanes that satisfy the equation, so the next step is to find the *optimized* hyperplane.

The optimal hyperplane expression, unfortunately, is usually a more complex function, and the optimization hyperplanes must be solved using the *Euler–Lagrange equation* and *Lagrange multipliers* of the *calculus of variations* to find the function that minimizes the margin extent under the constraint that the support vectors must be the vectors closest to the hyperplane (see the Appendix for the Euler–Lagrange equation and Lagrange multiplier). The support vector machine algorithms that perform the above operations are available on the Internet and generally free for users to try their hand at SVM.[2]

It can also be shown that the hyperplane expression is always concave so that there is only one global extremal, which avoids the vexing problem of local minima and maxima endemic to gradient descent in artificial neural networks.[3]

The marvelous aspect of Support Vector Machines is that in the kernelling transformation, the *kernel* can be simply represented by inner products in any dimension, and various transformation kernels can be tried, for example polynomial and exponential functions of the inner product vectors to find the optimal hyperplane and margins.

Simpler two-dimensional optimal line computations can also employ linear regression, but higher dimensional vector space may produce rather complex *organizing functions*.

Furthermore, although SVM can be employed in a "shallow" artificial neural network, there are fewer parameters than in typical ANNs to do the fitting.

Support vector machines have been used for text and handwriting classification, spam filtering, image recognition, colors classification, bioinformatics, and currently for handwritten digit recognition used in post office automation.

COMPUTATIONAL CHEMISTRY

Computational chemistry is the study of chemical compounds and their reactions with the goal of predicting the results of the reactions. The analysis includes thermal effects, temperature, pressure, concentration, chemical states and phases, catalysis, catalytic activity changes over time, solubility, impurities, large and small molecules, initial and boundary conditions, and so on; that is, a plethora of interacting chemical factors.

In addition, there are macro factors, for example in the petroleum industry, there are differences in the crude oil from different countries, the tanker ships and trucks, storage holds conditions, the shipping time, and the refining process, and there are ongoing exothermic chemical reactions that can foment chaotic chemical and physical changes, an enormous number of constantly changing conditions affecting the chemical and thermodynamical processes of a petroleum plant, and the results can further vary with just the size of the sample under study.

Typically, there are more than five or six simultaneous processes that must be *feature selected* from dozens of possibilities, and all optimized for efficient plant operation. The necessity of separating the immediately relevant factors from extraneous factors, as well as the considerable noise always generated in complex operations, altogether render the predictions of any model fraught with difficulty and uncertainty.

Experimentation and analysis of the data obtained, and an organizing hypothesis for making predictions for process engineering and new discoveries based on the analysis of those discoveries have been the subject matter of scientific and engineering research in chemistry, chemical engineering, materials science, environmental science, and pharmacology.

The analysis and predictions of chemical reactions have been done mostly by classical statistical methods such as regression analysis, but the major problem is that there are very few large petroleum plants that can provide data, and much of the data is proprietary and confidential, so the Law of Large Numberscannot be brought to bear.

Most of the reaction processes are complicated, nonlinear, multivariate, with copious noise, all of which have made useful and accurate computational chemical research models extremely difficult to design.

A linear relationship certainly makes the hypothetical regression organizing function simpler, but almost all the complicated chemical reaction processes of research interest are nonlinear, and linear regression considers *all* nonlinear data as *noise*, so a simplistic model has the risk of underfitting the data, while the use of more complicated functions with many terms and adjustable parameters to fit nonlinear data can result in too many adjustable parameterizations that will overfit the data.

The support vector machine's *support vector regression model* provides the simplicity of linear regression and can operate as an SVM with less data than traditional linear regression.

The use of multivariate regression, because of the large number of variables and their often very complicated nonlinear interactional relationships, typically results in a model having too many parameters, rendering its predictions and theoretical implementation results too variable to be useful.

If the organizing function is changed to a polynomial regression model based on the fact that any continuous function can be represented by a series of polynomials with an infinite number of terms (*Weierstrass' theorem*), appropriately truncating the series may adequately approximate the function if multi-term polynomials of various degree with different coefficients can better fit the data. But this requires many more terms and parameters, again bearing the very real risk of overfitting or underfitting the data with the loss of generalization for prediction and practical implementation.

Any continuous function can also be approximated by an artificial neural network, but ANNs are plagued by local minima, and also by the twin gremlins of underfitting and overfitting. In the case of computational chemistry, there are too many chemical physics parameters for good modeling and too little data to depend on the Law of Large Numbersto reach a useful generalizable result.

Because support vector machines can provide classification in a very high-dimensional hyperspace, there is in principle no limit to the number of factor segregations possible, and the SVM appears ideal for the highly complex computational chemistry of industrial processes having limited experimental data.

However, because of the myriad different conditions and operational factors in complex processes, some refinement of the data must be first employed to improve the efficacy of classification; for instance, *outlier deletion* where the data samples exhibiting large errors in supervised training can be *leave-one-out* (LOO) cross-validated, thereby parsing the data of at least some of the more obvious irrelevancies and noise.

Then SVM kernelling in the high-dimensional *feature space* of hyperplane vector space can be performed in a lower-dimensional *input space*.

In this way, nonlinear processes can be handled by SVM kernelling using the inner product of the data vectors involving a smaller number of adjustable parameters.

Support vector machines furthermore can treat both linear and nonlinear processes at once, so the problem of underfitting can be reduced, and if the number of parameters in the organizing hypotheses can be limited, hopefully there can be found a happy medium between under- and overfitting. The global extremal that can be found for the optimal hyperplane also makes the solution unique and the calculations therefore dependable.

In addition to computational chemistry, SVM can be used in scientific and engineering research where data is limited, and in basic chemical reactions, physical chemistry, quantum chemistry, atomic parameter stipulation, molecular structure, materials science, atomic orbital research, battery design, cancer diagnosis, gene structure, and chemical and materials product design.[4]

NOTES

1. "Distances" and "widths" in hyperspace are defined by vector inner products and are invariant in any dimension and in any coordinate system. For a marvelous mathematical derivation of SVM, see P. Winston 2014, *Artificial Intelligence*, MIT 6.034, online, which is an excellent demonstration of how to do mathematical physics.
2. For the sampling algorithms and examples, see Ludicky, L. and Torr, P.H., 2011, *Locally Linear Support Vector Machines*, Conference Paper, January, available at researchgate.com, pdf.
3. Ref, Fung, G., *Concave Minimization for Support Vector Machines Classifications* PowerPoint, research.cs.wisc.edu.
4. For an in-depth analysis of SVM in computational chemistry, refer to the author's cousin Chen, N.Y., *et al.* 2004, *Support Vector Machines in Chemistry*, World Scientific.

CHAPTER 21

Top-Down Speech Recognition

T HE FIRST TECHNICAL PROBLEM in natural language processing is to get spoken words into a machine for analysis. If you put your hand on your throat when you speak, you will find that there is a vibration, and longitudinal sound waves are being sent out, compressing the air in periodic puffs that in instances of unvoiced *stop sounds* (such as "p"), utterances will be analogously differentiated.

When the wave impinges a listener's eardrum, it will cause it to vibrate in step with the longitudinal impinging waveform, and an auditory nerve will send electrical signals proportional to the sound waveform through the outer, middle, and inner ear of the auditory canal, and then to the primary auditory cortex in the brain's temporal lobe, and finally to the parietal and frontal lobes of the human cerebral cortex of the brain for speech recognition.

Alexander Graham Bell, in 1876, used a diaphragm pressed against a bag of loose carbon particles, and ran a DC current through the bag. So when a sound wave vibrates the diaphragm, the distribution of carbon particles in the bag will change in step with the sound wave, thereby modulating the current and thus producing an electric current that represents the sound. Thus a "voice" current can be transmitted to a destination receiver in a wire which is wrapped around an electromagnet in contact with a diaphragm, the diaphragm will then vibrate in accord with the voice current and thus electronically reform the speech at the destination.

    DOI: 10.1201/9781003463542-21

Early in the 18th century, Joseph Fourier showed that his *Fourier Series* of sinusoidal functions could represent any waveform and his *Fourier Transform* could transform the series into a sound series of frequency as a function of time.

However, the calculation of a Fourier Transform requires a huge amount of adding and multiplication; for example, to transform a simple waveform with n points (typically about 1,000 points) requires adding the n points and multiplying their square n^2 which, before the computer, was an arduous, error-prone task that greatly limited its use.

However, about one hundred fifty years later in 1965, Princeton University and IBM produced the so-called *Fast Fourier Transform* (FFT) algorithm that greatly reduced the amount of adding and multiplying. The FFT would first divide the waveform into sample sets, combine the sets into points of a squared curve, and then recombine the sets so as to need only one multiplication step.

Needless to say, such a calculation decrease allowed the greater use of FFTs, but with the advent of computers, it would not make that much of a difference; however, the FFT became commonly used to analyze any waveform, including those in circuit design, signal processing, differential equations, image processing, and so on.

In machine voice *front-end processing*, the sound wave impinges the diaphragm of a microphone attached to an FFT analyzer, and an electronic graph is produced of the waveform on an oscilloscope, with different sounds producing differently shaped transverse waves.

An example of an electronic waveform is Amazon's "Happy Birthday" (Figure 21.1), which shows that different sounds have different transverse waveform shapes that start and stop as time progresses. Particularly the higher amplitude wave peaks called *formants* that form after stops and low amplitude sounds, and are generally vowels and stop/start sounds (like "b" and "p"), or other very distinct sounds.[1]

According to Fourier, no matter how complicated a wave is, it can be represented as an infinite *sum* of sine and cosine waves having different amplitudes a_n and frequencies ω as functions of time (if something can continue to ∞, it can be anything it wants to be),

$$f(t) = a_0 + \sum_{n=1}^{\infty}\left[a_n \cos(n\omega t) + b_n \sin(n\omega t)\right]$$

Happy Birthday

FIGURE 21.1 "Happy Birthday" speech waveform.

where a_0 is just the coefficient for $n = 0$, $a_0\cos(0\omega t) = a_0$, and there is no b_0 because for $n = 0$, $b_0\sin(0\omega t) = 0$. For $n \geq 1$, the a_n and b_n are the *Fourier coefficients*, which describe the "amounts" of cosine and sine of each frequency in the sound wave as represented by their amplitudes; the coefficients are given by,

$$a_0 = \frac{1}{T}\int_0^T f(t), a_n = \frac{2}{T}\int_0^T f(t)\cos(n\omega t)dt \; b_n = \frac{2}{T}\int_0^T f(t)\sin(n\omega t)dt$$

From the shape of the "Happy Birthday" waveform in Figure 21.1, it can be seen that a series of sine and cosine functions with different amplitudes and frequencies indeed looks like it has the potential to electronically represent speech.[2]

It turns out that the waveforms of *vowels* repeat themselves very clearly in sentences and because of their distinctive shape at different frequencies and relatively larger amplitudes, they have the greatest sound distinctions that can be uttered by humans, so a listener can easily distinguish them one from the other.

Human hearing is critically dependent on the perception of proportion, which is more distinctly manifested by a logarithmic rather than a linear scale. For example, the notes of the musical scale rise in pitch, and in different octaves the distance between them is perceived as about the same; that is, the distance from *do, re, mi* to the next octave *do, re, mi* when heard, seems to be monotonically increasing, but it is actually doubling in frequency (from 300 Hz to the next octave is 600 Hz).

The amplitudes of sound as a function of frequency are called *sound spectra*, which shows the location of formants that reveal the speech characteristics of loudness, pitch, intonation, and accent.

Natural speech of course is not strictly periodic (like sines and cosines), so the time length of the Fourier Series must extend to ∞ if the Fourier Series is to represent any and all waveforms of speech. Fortunately, this can be done by changing the variable in the equations above from time t to frequency f.

However, since frequency f is just the inverse of time period T, when the frequency interval goes to 0, the $1/T$ and $2/T$ coefficients of the Fourier coefficients will be undefined, but this is easily (and very usefully) remedied, since as the time period approaches infinity, the *frequency interval* in the spectrum of the wave goes to zero, so the discrete sound spectrum of amplitudes as functions of frequency conveniently changes from the *discrete* histogram peaks and valleys to become *smooth, continuous and differentiable* curves (like the waveform "Happy Birthday") are able to be represented by the elementary functions of sines and cosines.

Rewriting the Fourier coefficients in complex form and changing the $f(t)$ to $g(t)$ to avoid confusion with the f for frequency, and combining the Fourier Series and its coefficients equations, the Fourier transform and its inverse are,[3]

$$G(f) = \int_{-\infty}^{\infty} g(t) e^{-i2\pi ft} dt, \ g(t) = \int_{-\infty}^{\infty} G(f) e^{-i2\pi ft} df$$

where the $-\infty$ time in the lower limit of integration limit enables the past, but the negative frequencies from *minus-infinity* may give one pause, the above equations nevertheless give the frequencies and amplitudes to electronically model an auditory waveform.

The Fourier coefficient formulas represent the components of a waveform by extracting a single period of the wave and finding the area (integral) of that period for a given frequency, one frequency at a time.

Applied to a speech waveform, the inverse Fourier transform $g(t)$ is the integral over frequencies of the Fourier transform $G(f)$, which is an area that must be calculated. To do so, first the inverse Fourier Transform $g(t)$ can be taken as a function of discrete points in time, producing a plot of amplitude over time which can be read off of the speech waveform by the FFT analyzer; then multiplying those amplitudes by a sine wave just fitting into the range of time for which there are amplitude values, and with a period equal to the number of observed wave oscillations will produce a discrete, very fine, bar graph whose combined areas will be the area to match the integrations of the waveform.

The Fourier transform thus breaks up any waveform into its component simple waves, and renders the overall shape of the waveform recognizable from a sampled portion of it. The *discrete* Fourier transform was necessary for the FFT *digital* calculations by computers.

Through the next 40 years, the isolated word and speech pattern sound spectra were processed by *linear predictive coding* (LPC) and *dynamic programming*, and IBM and Bell Labs developed large-vocabulary, speaker-independent commercial speech recognition systems for use by computers and in telephony.

The technology generally comprised a *bank-of-filters front-end analyzer* for first separating the very different voice pitches (such as men from women), and producing a set of signals representing the energy of a sound in a given frequency band, thereby creating the sound spectra of an utterance.

With the graphical representation of any sound using electronic instruments, it was natural in the early days of speech recognition to use special purpose electronic hardware for top-down *acoustic–phonetic* speech recognition. For example, Japan's Radio Research Lab used a filter-bank spectrum analyzer with logic connecting each channel of the spectrum analyzer in a weighted manner to a vowel-decision circuit to recognize vowels.

The Russians first developed the critical time-aligning of a pair of *frequency warped utterances* (logarithmic perception of pitch) for *dynamic frequency warping*, and in America, RCA Labs modeled the nonuniformity of time scales in speech using a combination of both. Raj Reddy at Stanford pioneered continuous speech recognition by the dynamic tracking of *phonemes* (perceptively distinct units of sound that distinguish one word from another), incidentally first used for synthetic spoken moves in computer chess.

Linear predictive coding models the effects of the *glottal* (space between vocal cord folds) pulse representing sound intensity and pitch, the *vocal tract* (throat and mouth) resonances produce formants (distinctive frequency peaks), and the tongue, lips, and throat that produce the hisses and pops of a typical utterance, using time-dependent digital filters.

For the vowels, the formants are left–right symmetric about a central frequency on the logarithmic scale of frequencies, and from their spectral endpoints, one can find the maximum distance between formants, and the vowels thus may be classified as uttered "long" or "short" (such as the words "hear" and "heard"). Most electronic speech recognition systems, regardless of language, rely heavily on vowel recognition as a starting point.

LPC *signal source* front-end processing assumes that a given speech sample $s(t)$ at a given time t can be approximated by a linear combination of n past time speech samples $s(t - i)$ multiplied by *predictor coefficients* a_i, normalized by adding a gain factor G multiplied by a normalized signal excitation $u(t)$,

$$s(t) = \sum_{i=1}^{n} a_i s(t-i) + Gu(t).$$

Since the signal will change with time, the predictor coefficients at a given time must be estimated from a short segment of the speech signal occurring around that time, the estimate being performed at a rate of 0 –50 frames per second. The idea is to determine the set of predictor coefficients $\{a_k\}$ directly from the speech signal so that the spectral properties of a digital filter best match those of the speech waveform within the frame by minimizing the mean-squared error between the prediction and the speech sample for that frame.

The results of the filter-bank and LPC modeling are source-coded to convert the signals into a sequence of binary digits, and encoded in a series of vectors representing the time-varying spectral characteristics of the speech signal. This so-called *vector quantization* encodes an input vector into an integer index that can be associated through minimizing spectral distortion with a codebook of *reproduction vectors* that then can be used as a recognition preprocessor and/or training dataset for a speech recognition artificial neural network.[4]

In the acoustic–phonetic front-end recognizer, an input frame is matched to a reference set of features. Spectral features of compactness, gravity, stress, and flatness can be used as a reference to classify vowels with the decision as to the presence or absence of such features based on threshold values of acoustic parameters such as formant amplitudes, spectral band energy, and time duration. A vowel-decision tree then can be employed to sequentially test each proposition of the words in a speech sentence.

Acoustic–phonetic modeling using filter banks or discrete Fourier transforms for speech segmentation, labeling, and vowel and sound classification could extract features and largely identify individual words and some whole sentences, however because of the vagaries of strung-together spoken language, and the lack of a tuning mechanism to improve the recognition, the acoustic–phonetic approach by itself could not produce a generalizable automatic speech recognition system.

NOTES

1. Courtesy of Amazon screen shot.
2. One can easily form waveforms on a graphics calculator or personal computer by just adding the sine and cosine functions with different amplitudes and frequencies.
3. *Eulers formula* gives the relation between the exponential and trigonometric functions as $e^{i\theta}=\cos\theta + i\sin\theta$ so equations can be written in complex notation (meaning a "combination" of real and imaginary parts) which handles the square root of negative numbers, and is more compact, easier to manipulate, simplifies the use of zeros, and provides many a means to an end in mathematics and physics derivations.
4. Ref. Rabiner, L. and B.H. Juang 1993, *Fundamentals of Speech Recognition*, Prentice-Hall Signal Processing Series.

CHAPTER 22

Bottom-Up Speech Recognition

A FTER THE END OF World War II, the cold war between the Soviet Union and the United States began, and the mutual collection of intelligence became an issue of paramount importance. There was, however, a problem: neither side had a sufficient supply of translators to decipher the mountain of intelligence, often gathered at great risk by the spies. Both sides turned to the development of automatic Machine Translation (MT).

On the American side, in 1949, the mathematician Walter Weaver proposed a text translation machine; the idea seemed simple enough, given a document in Russian, a computer would text-recognize words by their spelling, look up the word in a stored Russian to English dictionary and construct sentences according to the rules of Russian and English grammar, also stored in memory.

The English language has at least 13 million words, and the number is growing every day with new slang, cultural, academic, technical, and social terms; but still plagued by the fact the *identification* of words is one thing, and the *meaning* of words in context, as well its effect on the meaning of the sentence itself, is quite another.

Natural language is fraught with ambiguity and inference, and the early MT attempts produced ridiculous translations such as the saying "the spirit is willing but the flesh is weak" being translated into Russian and then back into English as "the vodka is good but the meat is rotten".

DOI: 10.1201/9781003463542-22

The lack of progress together with machine translations such as "water goat" for the Russian "hydraulic ram" brought on ridicule culminating in the first AI Winter beginning in 1966 when the American National Research Council canceled all research support for automatic machine translation.

Any language translation must contend with sayings, usage, idioms, vernacular, slang, implication, innuendo, turns of phrases, puns, abbreviations, acronyms, and multiple meanings of the same word in different contexts, and other such unavoidable uncertainties and ambiguities.

For speech recognition, different speakers' varieties of accent, pronunciation, articulation, roughness, nasality, pitch, inflection, speed, timing, emotion, humor, sarcasm, and so on, all of which rendered accurate top-down machine translation of text and speech *literally* impossible.

The linguist Noam Chomsky's *Language Acquisition Device* (LAD) approach was based on *cognitive* language learning like a child's naturally learned speech recognition. This bottom-up approach was theoretically sound but research on cognitive understanding was sparse and the computational power of the computers of the 1960s limited, so Chomsky's LAD was ridiculed as just another amusing automatic speech recognition failure. But his approach was sound, and through computerization, just needed to be developed.

Since most words cannot be uniquely defined, words with similarities can be first grouped as elements of vectors, for example, as to synonyms (same meaning but different words) and homographs (same word pronounced in more than one way), homonyms (same sound and spelling but different meaning), homophones (same sound but different meaning), and heteronyms (same spelling but different sound and meaning) and then grouped as such, like the later-developed *synsets* of *WordNet*.

But the same word can have completely different meanings in context, such as the "hydraulic ram", so a contextual grouping must also be assembled, which understandably can be difficult.

Vectors, in addition to being able to contain ostensibly disparate elements that are similar in other respects, can also specify direction, and in-between distance by the inner product, possibly even to the extent of contextual meaning.

Furthermore, three words represented by x, y, and z could be related by their invariable distance d, from the Pythagorean theorem $x^2 + y^2 + z^2 = d^2$ for three dimensions and extendible to infinite dimensions. Thus, words can be classified as to their "closeness" in accord with the adage attributed to the linguist J.R. Firth,

You shall know a word by the company it keeps,

a truism that inculcates *similarities* and *context* into the understanding of a word.

Google's *Word2Vec* artificial neural network (ANN) first grouped words sharing common meaning into a vector, and then grouped the vector by closeness to other vectors by the inner product in a *sliding* window over the words dataset to recognize the words in accord with semantics, syntax, grammar, and context.

A *continuous bag of words* (CBOW) conversely uses an ANN with a softmax rectifier to predict which words from a window of surrounding words are most probably relevant in meaning and context.

A *skip-gram* uses a *center word* and ANN to predict the context in a surrounding *window* of context words, giving heavier weights to closer context words, and thereby helping to *fix* the word's meaning.

Natural language processing (NLP) employs two main statistical classification models, the *discriminative* and the *generative*. The discriminative model identifies the demarcation between classes and provides a label given an observation based on the conditional probability of an observable variable X and a target Y given an observable x,[1]

$$P\left(Y \mid X = x\right)$$

Examples are Decision Trees, Artificial Neural Networks, Linear Regression, Cross-Entropy Cost Function, and Support Vector Machines, all discussed previously.

The generative model estimates a *joint probability distribution* (here signified by "&"), and computes the conditional probability,

$$X \,\&\, Y, P\left(Y, X\right)$$

Examples are the Restricted Boltzmann Machine (RBM) and generative adversarial network (GAN) to be described in Chapter 24, Hidden Markov Model (HMM) and Gaussian Mixture Model (GMM), to be discussed in turn below; large language models (LLM) and generative pretrained transformers (GPT) will be introduced in Chapter 25.

WHAT IS THE NEXT WORD IN A SENTENCE?

In 1913, the Russian mathematician Andrei Markov took down his book-shelf copy of Alexander Pushkin's verse novel *Eugene Onegin*, not to read but to *deconstruct*, carefully writing out the first 20,000 letters and array-ing them in 20 × 20 matrices, counting the vowels, and meticulously look-ing for patterns revealing a mathematical structure of letters in verse that might be modeled.

Markov believed that the letters in a sequence of words are in a chain of causation. That is, in *Eugene Onegin*, the chance that a certain letter appears in sequence depends on the letter that came before it, and indeed his sample contained 43% vowels and 57% consonants, distributed as 1,104 vowel–vowel pairs, 3,827 consonant–consonant pairs, and 15,069 vowel–consonant and consonant–vowel pairs; therefore he believed that (certainly) *Eugene Onegin* was not a random distribution of letters, but might have a statistical character that could be mathematically modeled.[2]

An inner product calculation (such as *Word2Vec*) can be used to classify the input frame vector and a reference vowel feature vector using convo-lutional artificial neural network pattern recognition. The input speech is in the form of a time sequence of spectral vectors obtained from the front-end electronic spectral analyzers to form a *test pattern T* as a set of the spectral frame vectors over the duration of the speech,

$$T = \{t_1, t_2, t_3, \ldots, t_i\}$$

The test pattern T is compared with a set of reference patterns $\{R^j\}$ com-prising a sequence of spectral frames R^j,

$$R^j = \{r_1^j, r_2^j, r_3^j, \ldots, r_j^j\}.$$

Then minimizing the distance of T from each of the R^j will associate the input speech pattern with the reference *template*, and the global time alignment of the two patterns can be performed analytically using spec-tral distortion measurement techniques.

One spectral distortion measure is based on *frequency warping*, the human nonlinear, *logarithmic* perception of pitch, to model a wideband spectrum with a frequency resolution close to that of the human auditory system.[3]

Reference templates for training can be in the form of a nonrigid template or a statistical model. Templates are used in automatic speech recognizers, but even after undergoing training, their ability to adapt to different speakers, speaking styles, background acoustics, and electronic noise is limited, so they are typically used for very specific speech recognition tasks, such as recognition of a response to automated telephone answer requests, or as recognition preprocessors that can reduce the computational burden of a connected speech recognition artificial neural network.

Dynamic programming breaks down a complex problem, such as a long, spoken sentence, into subproblems, solving each of them and indexing the problem solutions based on their input parameters, and storing the solutions in a matrix. Thus, when the same speech problem is encountered, the solution matrix then can be looked up by means of its index, and the problem will not have to be solved again, thereby increasing computation speed and efficiency.

Dynamic programming has been widely used in operations research to solve sequential decision problems, and so it can also be used in speech recognition to account for past variations in speaking by using time alignment and normalization to predict the next word(s) in a sentence.

Summarizing, a front-end spectral analyzer measures short-time speech parameters sequentially, producing a sequence of spectral feature vectors. This speech pattern input is then compared with reference patterns, templates, or statistical models, and short-time and global spectral distortion (*dissimilarities*) are calculated using dynamic programming.

A subsequent step is to treat an utterance as a whole in a cognitive sense; that is, natural language *acoustic modeling* encodes the sound signal as a sequence of speech feature vectors whose frequencies instead of being scaled linearly, are logarithmically scaled (*warped*) to better model a human auditory perception that responds more acutely to logarithmic rather than linear proportions.

This so-called *perceptually warped frequency* is smoothed by the first- and second-time derivatives computed using differences of neighboring frames to capture the significant *temporal* influences in speech recognition.

These primarily electromechanical constructs using ANNs for pattern comparisons can provide *Automatic Speech Recognition* (ASR) for small vocabularies and limited speech variation, more general speech requires bottom-up deep artificial neural network learning.

In a conventional feed-forward artificial neural network, supervised learning is followed by reinforcement learning, but a *deep belief network* (DBN) first learns the probabilities of specific features from *unsupervised* training (the "belief"), and then while it undergoes *supervised* training, the DBN applies these *feature detectors* to classify the training set data, thereby learning from a head-start of "believed" features.

A speech recognition DBN thus can first act as a feed-forward network specifying the activation levels of the feature-specific neurons, and then by running the network feed-backwards, *generate* other features of the input data based on the learned speech.

In this way, it is ideal for recognizing the vagaries of natural language speech because the process is similar to a child naturally learning some basic words in unsupervised learning at home by listening to their parents' speech, and then going to elementary school for formal supervised learning of vocabulary and grammar; that is, before supervised learning, the child has some prior *beliefs* about the meaning of certain words and phrases and how to express them.

A DBN is therefore like a Restricted Boltzmann Machine (RBM) acting on probabilities to reveal latent factors in speech from *reconstructed probabilities* that will help it to recognize the implicit meaning of individual speech from context, usage, and all the other vagaries of spoken communication, and then by gradient-descent and backpropagation, the words and phrases can be collaboratively filtered to include the *inferences* so common to natural speech, and thus move closer to *Natural Language Processing* (NLP).

In the modern *Hidden Markov Model* (HMM), a Markov Chain using *Gaussian Mixture Models* (GMMs) is employed to generate probability distributions for the acoustic vector sequences produced by front-end electronic acoustics, then individual Gaussians (the mixture) generate the variables in the multiple dimensions required by speech recognition to form the distribution of vectors in a matrix.

However, context-dependent models clearly require a great deal of speech data to be accurate, so for more efficient use of the data, the HMM is divided into sub-HMMs for each *triphone* (sequence of three phonemes) and HMM alpha–beta pruned decision trees to associate different speech situations.

In human speech, sequences of utterances are meant to represent isolated words or phonemes, but a speech recognizer does not know *a priori* what the words are meant to mean; that is, the actual meanings are embedded in states *hidden* from the speech recognizer, but since the utterance is heard, there are *observables* (*hearables*?) from which the intended meaning of the words may be *inferred* based on the probabilities of occurrence.

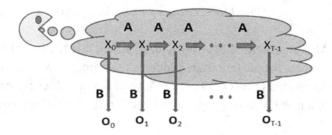

FIGURE 22.1   Hidden Markov Model.

A simple Hidden Markov Model has state transition probabilities A and hidden state sequences X_i in a cloud of uncertainty, below which are an observation probabilities matrix B and the observation sequences O_i, for the total sequence time T, as schematically illustrated in Figure 22.1.

Using HMM, the highest probability word at each point of the sequence can be chosen by summing the probabilities that $word_1$ is in the first position with $word_2$ having a probability of $(1 - word_1)$. Repeating this for each element in the sequence gives the probabilities $P(word_1)$ and $P(word_2)$ for each element in the sequence, thus obtaining the most probable sequence for the two words in the utterance.[4]

Generalizing to multiple words, phonemes, sentences, phrases, and so on will increase the dimensions of the state transition probability matrices and add complexity to the Hidden Markov Model, but the general idea is as just described.

When each element in the principal diagonal of the matrix is a *variance* of one of the other elements, meaning that words are very close to each other in some manner such as described above (a better word would be "variant" as "derived from" or "closeness"), the matrix is *diagonally covariant*, and such matrices thus represent speech relationships that are used to produce the *joint probability distributions* needed to associate the context- and time-dependent aspects of natural language speech.

Since there are many more possible observables and their sequences than labeled training data, a probability distribution is calculated for each time step for the alignment of the input speech sequence with the training data. The state transition matrix A has probability elements a_{ij} where

$$a_{ij} = P(state\ q_j\ at\ t + 1 | state\ q_i\ at\ t)$$

and the observation probabilities matrix B has elements $b_j(k)$ where

$$b_j(k) = P(observation\ k\ at\ t|state\ q_j\ at\ t).$$

An HMM is represented by $\lambda = (A, B, \pi)$ where π is the initial state distribution. For a simple four-state hidden state sequence $X = (x_0, x_1, x_2, x_3)$ with observations $O = (O_0, O_1, O_2, O_3)$, with scalar values for example a vowel (0), a hiss (1), and a pop (2) observed in the four-state sequence as (0, 1, 0, 2).

The probability of state sequence X is,

$$P(X,O) = \pi_{x_0} b_{x_0}(O_0) a_{x_0,x_1} b_{x_1}(O_1) a_{x_1,x_2} b_{x_2}(O_2) a_{x_2,x_3} b_{x_3}(O_3).$$

The probability of any sequence of utterances can be thus calculated, and for each state sequence, for example the given observation sequence (0, 1, 0, 2). Then using dynamic programming, the state sequence with highest probability will be the best choice of word sequence in the given phrase.

The most basic probability distribution is just a random Bayesian distribution, and the *Bayes' formula* is used for the probability based on further data.

$$P(A\ B) = \frac{P(B\ A)P(A)}{P(B)}$$

As in all statistical probabilistic models, the theory can be mathematically dense, but the principles are relatively easy to understand and the computations can be done online using packaged software.[5]

The Gaussian Mixture Models (GMMs) group data points into clusters within which they are Gaussian (normally) distributed. GMMs have been employed to model the spectral representation of a sound wave, and can classify groups of data for the representation of phrases and sentences. *Factor analysis* represents each data point as a weighted linear function of latent inferences in the data, thereby introducing sophisticated nuance into automatic speech recognition.

RECURRENT NEURAL NETWORKS

Speech recognition in general is critically dependent on *time and sequence*, so the HMM GMM word probabilities are typically forward-fed to a so-called *Recurrent Neural Network* (RNN), a *transformer* whose activations are time-dependent employing *Connectionist Temporal Classification* (CTC) that can be used to train an RNN on time and sequence employing *Long Short-Term Memory Networks* (LSTM).

The total ASR system can be further refined by reinforcement and self-supervised learning by running against itself for self-improvement for *a posteriori* parameter optimization producing greater accuracy.

In typical feed-forward neural networks, a single input layer completely determines a *static* synaptic activation pattern throughout the other neuron layers. In a *Recurrent Neural Network* (RNN), the neurons can be controlled to only fire for a limited time duration, so the activations of succeeding neurons in the synaptic pattern will be influenced and such influence can be carried on to succeeding neuron synaptic patterns, giving the RNN a temporal sensing capability. A given neuron may even respond to its own earlier activation to connect an association, thus forming a *temporally controlled cascade* of activation that can manifest a speech pattern based on time-based preceding patterns of activation.

In this way it can be seen that the all-important timing of speech can be represented by timed (*volatile*) neural firing patterns.

Apart from any hand-engineered associations germane to the particular speech recognition implementation, for example the close words "checking" and "balance" in bank telephony speech, the recurrence is typically performed by inner products of feature vectors that provide the scalar correlation between the speech feature vectors.

The recurrent neural network thus can respond to stimulations depending on the prior presence of close feature vectors that account for related inferences, or the prior absence of signals to indicate lack of relation, thereby providing more accurate probabilities of later word feature occurrence.

That is, RNNs can connect relatable previous information to present meaning based on the closeness of the feature vectors. For example, the phrase, "I grew up in France …" earlier in the text or speech can imply a *recurrence* with the later occurring phrase "I speak fluent _____", where the earlier occurring word "France" generates a high inferential probability (manifested by the closeness of word feature vectors) that the blank should be the word "French", even if the word in the blank space is garbled and there are many words and pauses in-between. The RNN thus has the ability to provide a word through inferential association of earlier speech by the closeness of word feature vectors even if the word in text is misspelled, unclearly written, or mumbled in speech.

On the other hand, if the later occurring phrase is "I *also* speak fluent _____", then the association should not be made because of the word "also" implying another language and the prior word "France" can be deactivated in this instance, while other countries' names can be activated, possibly through prior occurrence.

Another example is *uptalk* (voice lifting in pitch and inflection at the end of a sentence) where a statement may be mistaken for a question, and can be determined by earlier instances of similar uptalk occurrence. In this way, the RNN can produce a *word scores* matrix in accord with the context of the speech at issue and a speaker's particular speech intonation.

If the first part of a subject occurred at the beginning of the speech, and the last part near the end, and there is a considerable span between the related phrases, the RNN's sequential activation can place the first part in a *stored state*, which can be under the control of the RNN as a *controlled state* with time delay and feedback loop capability for reactivation as needed in the event of the last part appearance.

If the earlier occurring feature is more extensively referred to later, then the stored controlled state may be recorded entirely in another network or data graph that incorporates time delays and can be fed back to the RNN.

Theoretically, the information can propagate arbitrarily far down the sequence if it is still encoded, but the vanishing gradient problem can leave the model's state at the end of a long sequence lacking information about the preceding information. The dependence of information on the results of previous information computations makes it more difficult to parallel process using GPUs.

Recurrent neural networks can use all the deep neural network convolutions, regularizations, and other feed-forward techniques to more accurately perform speech and text recognition. These techniques are particularly useful for any task where memory of past events, thoughts, and features are significant for present processing.

Marcel Proust's *Remembrances of Things Past* is a literary example of a human RNN describing impressions of long-past events to the minutest detail.

Since recurrent neural network backpropagation is performed not only through event layers, but also through temporal layers, the problems of vanishing and halting gradient descents can result in slower and sometimes even null learning.

This problem is addressed by controlling the gradient-descent instability by limiting the backpropagation through the use of *gated states* or *gated memory*, conjunctively termed *gated recurrent units* (GRUs), in an incongruously named *Long Short-Term Memory Network* (LSTM). This name only makes sense in the context of the Automatic Speech Recognition (ASR) gated recurrent unit regime.

The acronym-laden ASR LSTM GRU can add or block information by means of a layer of three sigmoid function gates that lets information through based on its greyscale. So if an earlier word whose occurrence is helpful in predicting a current word or future word, its activation will be passed on by an "ON" (*1*) as a long-term open-gated recurrent word, while other long-ago words that are not helpful (meaning word vector features are not close to the present speech), in the short-term speech processing will be logic gate blocked (*forgotten*) by the GRU as an "OFF" (*0*) state.

This explains the "long" as a long distance or time away (that is, not a "close" word vector), and "short-term" as just needed for this particular short real-time prediction of the meaning of the word or phrase in question. LSTM RNNs therefore can learn the long-term dependencies for the immediate needs of speech recognition.

A *Connectionist Temporal Classification* (CTC) is used for training recurrent neural networks employing LSTM to do sequences where the timing is variable and can be reinforcement learned. A CTC network trains the RNN by taking the word scores matrix from dynamic programming and then infers the speech or text pattern from the state transition probabilities matrices.

The neurons in recurrent neural networks thus are *dynamic*, like a biological brain, and therefore CTCs are effective for modeling processes that change with time in a sequential manner, for example, cursive (connected) handwriting recognition and natural language speech. Even Audio-Visual Linked Speech Recognition (AVSR) models that link sound with vision observables, such as hand gestures and lip-reading, have been developed.[6]

In summary, most automatic speech recognition systems represent speech as a sequence of perceptually warped feature vectors with smoothed differences of neighboring frames from first- and second-time derivatives. The probabilities of feature vector sequences are modeled by Hidden Markov Models with Gaussian Mixture Models (GMMs) where the HMM is constructed from sub-HMMs for each triphone, and the individual Gaussians are all diagonally covariant matrices.

Clustering the HMM states using alpha–beta pruned decision trees can produce desired parameter-tying. The HMM–GMM word probabilities typically employ Recurrent Neural Networks (RNN) which in turn use Connectionist Temporal Classification (CTC) and Long Short-Term Memory Networks (LSTM) to process time and sequence, and reinforcement and self-supervised learning can produce parameter optimization for greater speech recognition accuracy.

Automatic speech recognition follows the arc of natural language, fraught as it is with uncertainty, ambiguity, and inference; natural language processing thus is an exceedingly complex top-down signal processing endeavor that requires a succession of bottom-up artificial neural networks and techniques to succeed.

All the network computations described above can be performed using Python, *PyTorch* framework, and the *TensorFlow* platform, employing a *Computation Graph* that allows code reusability and extension available to any interested programmer.

In addition to setting new records for accurate text and speech recognition, a recurrent neural network learned the character-by-character sequence used in the high-level computer program language *Python*, and in a sequential, dynamic way learned how to write computer programs in *Python*, threatening the very livelihood of computer programmers worldwide.

But more distressing than the obsolescence of computer programmers might be the total taking over of not only all textual writing tasks, including translation; and generating images, music and …

NOTES

1. The vertical line | denotes a conditional probability; that is P(A|B) means the probability that A occurs given that B has occurred.
2. Rf. Markov, A., cited in O. Schwartz, *IEEE Spectrum*, Nov. 12, 2019.
3. For details see Rabiner, L. and B.H. Juang 1993, *Fundamentals of Speech Recognition*, Prentice-Hall Signal Processing Series.
4. M. Stamp 2018, *A Revealing Introduction to Hidden Markov Models*, October 17, online.
5. Ref. Rabiner, L. and B.H. Juang 1993, *Fundamentals of Speech Recognition*, Chapter 6, Prentice-Hall Signal Processing Series.
6. For example, Zhang, Y. "Speech Recognition Using Deep Learning Algorithms", cs229stanford.edu and *Audio Visual Speech Recognition and Segmentation Based on DBN*, researchgate.net.

CHAPTER **23**

Speech Synthesis

T HE EARLIEST SPEECH SYNTHESIZERS were developed hundreds of
years ago by scientific and engineering luminaries such as Albertus
Magnus, Roger Bacon, and Charles Wheatstone. These *articulatory synthesis* systems mechanically model the human vocal tract, with vocal fold
biomechanics, glottal aerodynamics, and acoustic wave propagation in
the biomechanical bronchia, trachea, nasal, and oral cavities of *mechanical talking heads* powered by puffs of air from bellows.

More recently, Bell Labs developed a voice codec (*Vocoder*) called the
Voder for electricity-generated telephone speech. As diagrammed in
Figure 23.1, an operator creates vowels by depressing a wrist bar producing nasal buzz tones. Consonants are generated by a white noise tube producing a hiss, with a foot pedal to control pitch, and the explosive "*p*"
and "*d*" and the affricative "*j*" and "*ch*" activated by *spectrum keys* that
select from ten band-pass filters for emulation of these basic sounds to
form combinations of words that are transmitted to a speaker for demonstrations of synthetic speech.[1]

The Voder was displayed at the 1939 New York World's Fair with the
greeting, "Good afternoon, radio audience". Needless to say, intelligible
speech generation required no little training and great skill of the operator.

In the late 1940s, speech synthesis pioneer Franklin Cooper developed
the *Pattern Playback* machine that converted sound spectra spectrographs
of patterns of speech into audible synthesized speech, and in 1968, Bell
Labs employed the ubiquitous IBM 704 computer to synthesize "Daisy
Bell", a song that was subsequently played by the computer HAL in Arthur
C. Clarke's screenplay for the film "*2001 A Space Odyssey*".

DOI: 10.1201/9781003463542-23

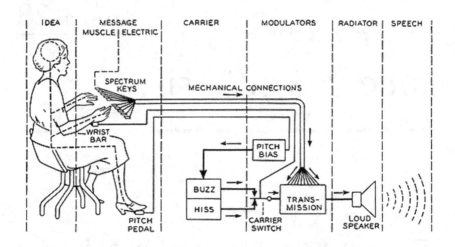

FIGURE 23.1   Electromechanical synthetic speech Vocoder.

With the advent of the computer, the rudimentary air, electricity, and sound spectra spectrograph sources of synthesized speech could be replaced by computer software and refined for greater verisimilitude.

Linear predictive coding (LPC) for speech synthesis was developed at Nagoya University in Japan, electronic talking heads were developed in America and Japan using digital synthesis to produce articulatory speech firmware, and Texas Instruments' LPC microprocessors were used in its *Speak and Tell* toys popular in the late 1970s.

Japan's NTT in 1975 developed the *Line Spectral Pairs* (LSP) that mathematically pair the LPC equation predictor coefficients a_i for improved stability and resonance (see Chapter 21), and these *LPC filters* were used to more closely match electronic speech waveforms. The LSP technique was subsequently adopted in 1990 as the international speech coding standard for mobile telephony and the Internet.[2]

With more accuracy, speech synthesis could be extended to more general uses. The first text-to-speech (TTS) system was developed in 1975 by Italy's *Centro Studi e Laboratori Telecommunicazioni* (CSELT) with the *Multichannel Speaking Automation* (MUSA) dedicated computer and diphone-synthesis software. MUSA was able to read aloud and sing Italian songs from printed text. Later, Bell Labs, MIT, and Digital Equipment Corporation in the 1980s developed the TTS *DECtalk* natural language processing (NLP) computer for multilingual text-to-speech synthesis.

The basic TTS process is for a *front-end* processor to convert written text to a phonemic representation by first translating numbers and abbreviations into the equivalent written words (*text normalization* or *tokenization*), distinguish homographs, for example whether "read" should be voice synthesized as "red" or "reed" (determined by *part-of-speech tagging*), assigning phonetic transcriptions to each word, and segmenting the text into word, phrase, clause, and sentence *prosodic* (pitch contour and phoneme duration) units, a *back end processor* then performs prosody prediction and generates waveforms for discrete to continuous synthesized speech as shown in the *continuous back end processing* flowchart of Figure 23.2.

In the so-called *concatenative synthesis*, segments of recorded speech in the form of electronic waveforms are strung together, and although an individual string may sound quite natural, there are noticeable pauses between strings that everyone has experienced in early automated

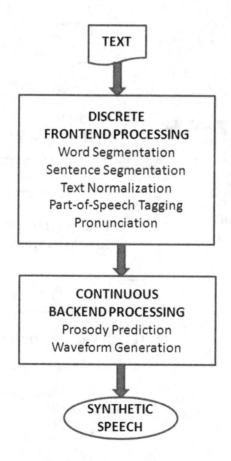

FIGURE 23.2   Continuous backend processing.

telephone answering services (Figure 23.3). These are typical "pipeline" systems flowcharts.

These models, together with *high-level audio synthesis* (such as Google's *Tacotron*) and deep learning network *waveform synthesis* (DeepMind's *WaveNet*), comprise a complete text-to-speech engine.

A *domain-specific synthesis* employs a set of prerecorded words and phrases that are natural sounding, but limited to the domain, such as for the early talking dolls, scales, and clocks. In *diphone synthesis*, one sample of a larger set of sound-to-sound transition speech is used for each word, and linear predictive coding or discrete cosine Fourier transforms can be applied to the diphones to provide sentence prosody, but because of the database limitation to one sound per word, the speech cannot help but sound robotic.

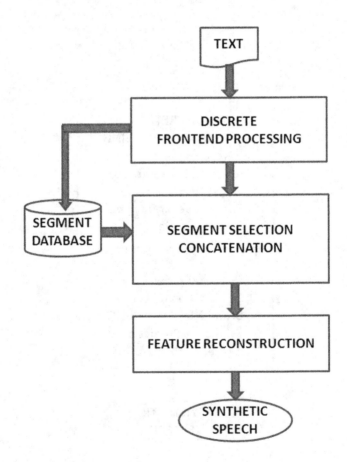

FIGURE 23.3 Concatenative synthesis.

In *formant synthesis*, the synthesized speech is wholly formed from electronic signal processing of the frequency sound spectrum amplitude peaks (formants), and then adding sine waves together (*additive synthesis*), or using mathematical physics models with Fourier transforms to create whole new waveforms. Although capable of more generalized synthesis, this not unpredictably produces rather electronic-sounding speech.

Because of the sophistication of the digital signal processing, digitally synthesized speech can provide articulation that eliminates the unnatural pauses of concatenative and formant synthesis models, and can produce quite natural speech, and modulations of prosody and intonation can produce emotion and tone, useful particularly for humanoid robots.

Synthesized speech in response to human speech commands or questions of course first requires the recognition of the text or spoken input and then the choice of an appropriate response from the files in its synthetic speech database.

In *unit selection synthesis*, recorded utterance waveforms are divided into phones (any speech sound), diphones, half-phones, syllables, and morphemes (minimal grammatical units of a language that cannot be broken down into smaller independent grammatical parts) to form the files in the database.

To construct a sentence, an index of these *units* is based on the segmentation and acoustic parameters of pitch, duration, position, and neighboring phones so that related words and phrases can be classified, and synthetic speech of whole sentences can be constructed by using a weighted decision tree of the most probable chain of tokens derived from the database.

Artificially intelligent robots clearly require speech synthesis. For example, Hidden Markov Models can include complete dictionaries that can be searched for pronunciation based on spelling or rules or combinations thereof to handle part-of-speech tagging. Deep neural networks can train the model from recorded speech datasets to produce natural-sounding words, and HMMs can model the sound spectrum, pitch, and duration of word waveforms probabilistically to form natural language-sounding sentences.

Then just as in speech recognition systems, recurrent neural networks, LSTMs, CTCs, and other networks and models employing supervised and self-supervised learning can be employed to refine the text-to-synthetic speech, synthetic speech-to-text, and text-to-synthetic speech.

These technologies are used in the audio-editing software-generating tools that can be trained to produce synthesized speech that closely mimics a particular speaker by taking a voice sample, and generating characteristic speech that through the employment of speech recognition inferences, can even include phonemes that were not in the training data set.

More recent speech synthesis engines such as OpenAI's GPT-3 neural network *Whisper* is an end-to-end encoder/decoder transformer where input audio is parsed into 30-second samples, encoded into a sound's short-term log-power spectrum based on a linear cosine transform on a nonlinear *mel scale* for speech-to-text. It can translate text-to-speech in more than 50 languages and transcribe 99 languages to English (and additionally can lower communication sounds for privacy, particularly downloaded for mobile phones).[3]

Speech-to-speech translation (S2ST) synthesis obviously is not necessary between humans speaking the same language, but is critical for automatic translation and the development of humanoid robots' speech and translation. LLMs and GPT-n can perform S2ST, but not necessarily better than the methods described above.

The speech recognition module recognizes the speech and compares it with a phonological model. It is then converted into a computerized string of words (not word-by-word) using a dictionary and grammar LLM and transmits to a speech synthesis model such as shown in Figures 23.2 and 23.3 that textually estimates pronunciation and intonation from the string of words, and finally waveforms matching the text are chosen for synthetic speech output.

Speech-to-speech systems have the problems of incoherence, slang, sarcasm, cynicism, wordplay, and so on, but on the other hand, is simpler than TTS because of the smaller vocabulary and simpler structure of general speech.

AUTOMATIC TRANSLATION

Search engine optimization (SEO) can already translate 204 languages with about 80% accuracy using a mobile phone app, and *statistical machine translation* (SMT) employing statistical correlations matching from available data, but the database requires at least several million words, and is not more accurate than SEO.

From this, the LLM GPT-n with the entire Internet and Cloud as vocabulary should do better, but its COMET score is still below specialized neural network machine translation (NMT).[4]

Google Translate uses several million machine translation examples, Amazon Translate uses AWS' transfer learning, allowing the translation of different local dialects and colloquialisms, Microsoft's Bing, Germany's DeepL, and China's TenCent Machine Translation (TMT) are just some of the many voice translation models.

Natural language synthetic speech, like all technological innovations, can be and have been abused, for instance by the unethical putting of words into the mouths of public figures in commercials, parodies, or for political gain.

However, one public figure's thoughts were synthesized not for nefarious, comedic, or political gain, but rather for exposition of the deepest mysteries of the Universe. The late renowned theoretical physicist Stephen Hawking's eerily robotic voice at first used DECtalk, but this required him to type the words for TTS synthesis, and he was increasingly unable to do so as his hand muscles degenerated from ALS.

In extraordinary displays of sensor technology and artificial intelligence, subsequent speech synthesis systems developed by Intel, SpeechPlus, and advised by Hawking's graduate assistants, followed twitches in his cheek muscle to predict word selection from a deep neural network trained on his books, papers, and speeches; for example, he had merely to twitch his cheek muscle in a particular way for the word "the" and a recurrent neural network immediately produced the contextually concatenated inferred words "black hole".

Stephen Hawking's AI-deduced speech that allowed him to live a longer and extremely productive life for the good of science, ironically also included words on the dangers of the looming *AI Singularity*,[5]

> *I fear that AI may replace humans altogether. If people design computer viruses, someone will design AI that improves and replicates itself. This will be a new form of life that outperforms humans.*
>
> *It will either be the best thing that's ever happened to us, or it will be the worst thing. If we are not careful, it very well may be the last thing.*

This pronouncement was made more so dramatic precisely because it was itself robotically generated.

Hawking's warnings indeed might materialize, but for now the AI Singularity, at least for speech synthesis, has not yet come to pass, for no one would mistake his synthesized words for natural human speech. This elucidates the fact that although synthesized speech has progressed to the point of *intelligibility* and *generality*, mostly because of awkward pauses and strange syllabic emphasis, synthetic speech is seemingly forever hampered by a lack of *naturalness*.

In IBM's Project Debater, argument mining used *knowledge graphs* that gathered information from many sources (such as *Wikipedia* and the *CIA World Factbook*) comprising billions of facts that were relationally organized in the so-called *knowledge boxes* to assess controversies and dilemmas and model the commonalities and discrepancies of the information data, one of the early LLMs.[6]

Responses were transcribed and text-to-speech synthesized so that claims, rebuttals, and arguments were offered in continuous and inflected speech for cogent, intelligible, and persuasive arguments, abetted at times with incongruous robot-humor. For this, IBM developed TTS algorithms employing expressive synthetic speech models with predictable phrase breaks and word- and sentence-emphasis.

All that this historic debate lacked was a curtain hiding the debaters from the audience's view, for the debate could have been a true *Turing Test* if after the debate, the debate host had asked the audience to distinguish the human from the machine, and if they could not, Miss Debater would have established the knowledge and intelligence at least *equivalent* to a human, and particularly not any human, but an accomplished champion human debater.

However, a clear giveaway in the Turing Test would have been the speech synthesizer at times peculiar enunciation of obscure technical terms, foreign words, and unnatural pauses, even if delivered with colloquial certitude.

And this, it may be surmised, was a factor in Miss Debater's loss to Harish Natarajan. Although the female voice was used to soothe fears of robots, because of its electronic synthesis, it remained irredeemably robotic, and while robot-humor may put one at ease and amuse, it may also dismay as well.

Furthermore, Miss Debater's ominously challenging opening statement did her little good in winning over an audience composed entirely of humans; portents of unalloyed robot superiority hubris can easily diminish any goodwill from the human victims.

Just as in the case of the audience's animosity displayed against Deep Blue, the human audience likely subconsciously sided with the human, revealing a deep-rooted psychic fear of machines besting humans.

A machine won all the *objectively* scored challenges, the only contest that the human won was the *subjectively* judged debate. Perhaps an audience of more robot-appreciative geeks, computer scientists, or robots themselves would have voted for Miss Debater.

In retrospect, Harish Natarajan won at least in part because Miss Debater's speech no matter how well-synthesized would nevertheless create some cognitive dissonance, while the urbane and unaffected Natarajan's debating delivery no doubt produced an attractive resonance that no synthetic voice could then, or perhaps ever, match. After the debate, Harish Natarajan had said that "I felt I had an advantage because I was not a machine, I was a human", and what gave away Miss Debater was her, steadily improving, but still robotic synthetic speech.

Completely identical to human speech synthesis may also just be a fool's errand, completely mimicking human colloquial speech may be impossible, and even more may not be necessary or desirable, since the false impersonation has its dangers. Why can't humans just accept robot-accented speech, and know that it is a robot that is speaking and not a human, just like recognizing the accent of a foreigner?

NOTES

1. Figure from 1928 patent by Homer Dudley is in the public domain and from Wikipedia Commons. It was used to encrypt messages during World War II.
2. For the LPC equation, refer to Chapter 21.
3. The Mel Scale is devised so that sounds of equal distance from each other on the Mel Scale "sound" to humans as if they were equal in distance from one another.
4. COMET Crosslingual Optimized Metric for Evaluation of Translation.
5. Quote from *Economictimes,* Indiatimes.com, March 14, 2013.
6. Knowledge graphs and boxes are used by Google for information searches and in *Google Assistant* and *Google Home* to answer spoken questions. Google's *knowledge vault* automatically gathers and collates data from text, but suffers from lack of information attribution (which in principle can be done by automated search as well). Google's knowledge vault's technical details are proprietary.

CHAPTER 24

RBMs, GANs, and LFCF

A RESTRICTED BOLTZMANN MACHINE (RBM) is an early artificial neural network with only an input (*visible*) layer composed of vectors *v*, one *hidden layer* composed of vectors *h*, and no output layer. There are no node connections among the artificial neurons *within* the layers (hence the "restricted" adjective in its name).

Whereas regression models estimate a continuous dependent variable based on the independent variable data input, and ANN extracts features from the data, an RBM iteratively reconstructs the *probability distribution* from unstructured input data, and thus is performing *generative learning*.[1]

Probability distributions are based on the probabilities of a set of outcomes. For example, in rolling dice, out of a total of 36 possibilities, the probability of a lucky "7" is six times higher than the probability of snake eyes "2", since there are six ways of the dice adding to 7 and one for a snake eyes "2". The probability distribution will be revealed as a Normal distribution curve with the "7" at the peak and "2" and double six "12" on the wings of the Gaussian bell curve.

Another example is in speech recognition, *e*, *t*, and *a* are the most commonly used letters in the English language, while in Icelandic, the most common letters are *a*, *r*, and *n*, so the probability distribution curves for the alphabet letter occurrence in English and Icelandic speech processing are substantially different.

RBMs are based on Boltzmann thermodynamics physics and Gibbs *free energy* chemistry to determine distribution probabilities; the RBM algorithm is a Markov Chain random field of connected nodes where the *joint*

DOI: 10.1201/9781003463542-24

probability of the neuron activations in the layer vectors "*h* given *v*" and "*v* given *h*" can be represented by that free energy, which is a measure of the stability of the probability distribution; that is, the less free energy in the system, the more stable the system, meaning for artificial intelligence, the closer it is to the ground truth of the training dataset.

The *Gibbs free energy* of physical chemistry is a measure of the thermodynamic potential of a state of matter. For example, on Earth, H_2O has three phases: liquid, solid, and vapor. At room temperature and atmospheric pressure, although there is some water vapor in the air (the humidity), the liquid state of H_2O has the lowest free energy and is the most stable of the three states, so as ice cubes in an ice tray on the kitchen table melt, free energy is released in a phase transition of ice to liquid water, and as liquid water in the air evaporates (dehumidifies), free energy is released in the phase transition of liquid to vapor, and the free energy of the system decreases.

In other words, under standard temperature and pressure (STP) conditions, liquid water is the most stable state of H_2O, meaning that it has the highest probability compared to the other states, which exist but have more free energy and are less stable and therefore less probable.

The ice on the North and South Poles of the Earth is changing states by melting from the cold, originally more stable, solid ice to water because of global warming caused primarily by the emission of increasing levels of carbon dioxide vapor forming a gaseous dome blocking the escape of infrared heat from the Earth.

If say on another planet, the equilibrium temperature is considerably higher than 100°C or considerably lower than 0°C, or the pressure is not near atmospheric, any chemical compound having a vapor, liquid, and a solid state could have less free energy than the other states and be more stable, meaning that if conditions change, the ground truth probability distribution will be different. For example, a thick methane vapor cover and enormous sea of liquid methane with crystals of solid methane below on Saturn's largest moon Titan, which apparently has an Titanian STP environment conducive to a CH_4 triple point.[2]

An RBM, starting from a random initial distribution, compares it with the distribution of the input data; the difference is just the free energy of the candidate distribution, so just as in other artificial neural networks, minimizing that free energy will cause the RBM-generated probability distribution to converge to the probability distribution of the ground truth input data, and reveal the latent inferences hidden therein.

The Gibbs free energy of a pair of Boolean vectors (v, h) representing the visible v and hidden h layers is given by the RBM *energy function*,

$$E(v,h) = \sum_i a_i v_i + \sum_j b_j h_j + \sum_{i,j} v_i h_j w_{ij}$$

where a_i is the activation energy of the ith neuron, v_i the binary state of the neurons in the visible input layer, and h_j the binary state of the neurons in the hidden layer; the b_j are the elements of the bias vectors, one for each layer, and w_{ij} the weights elements of the weight matrix W.

The thermodynamic probability P_i of the ith state of a system having an energy E_i at temperature T is given by the well-known Boltzmann Distribution (hence the name of the RBM machine),

$$P_i = \frac{e^{-E_i/k_B T}}{\sum_{j=1}^{M} e^{-E_j/k_B T}} = \frac{e^{-E_i/k_B T}}{Z}$$

where M is the number of all possible states in the system, k_B is the Boltzmann constant (relating temperature and energy), and Z is the *canonical partition function* that normalizes the equation to values between 0 and 1 as required for probability calculations.

For an RBM, the joint probability $P(v, h)$ of v given h, and of h given v, depends on the RBM energy function $E(v, h)$, and is given by,

$$P(v,h) = \frac{e^{-E(v,h)}}{\sum_v^M \sum_h^M e^{-E(v,h)}} = \frac{e^{-E(v,h)}}{Z_{rbm}}$$

where Z_{rbm} is the canonical partition function sum over all possible pairs of visible and hidden states.

At a given point in time, the RBM-generated probability distribution is in accord with the RBM energy function $E(v, h)$, which energy is determined by the parameterized activation levels of neurons in the visible and hidden layers. In feed-forward mode, the RBM thus is acting as an *autoencoder*.[3]

The calculation of the possible probabilities of all the states of v and h is prohibitively dense, so simpler *conditional joint probabilities* of h given v and v given h are performed by the Π operator multiplication over the index i,[4]

$$p(hv) = \prod_i p(h_i \# v)$$

$$p(vh) = \prod_i p(v_i \# h)$$

Since each neuron activation level by itself is binary, it can only be *1* or *0*, the weight and bias parameterization are factors, and of course are effective only for the case that the neuron activation level is *1* and not *0*. For given neuron activation levels of the visible layer *v*, the probability that a single neuron in the hidden layer *h* is an activated binary 1 with level adjusted by the shared weights w_{ij} modulating the visual layer neurons v_i is,

$$p(h_j = 1v) = \frac{1}{1 + e^{-(b_j + \sum_i v_i w_{ij})}} = \sigma(b_j + \sum_i v_i w_{ij}) \quad (\text{Eqn.1})$$

where σ is the sigmoid function. There are two biases b_j in the RBM auto-encoding, the *hidden layer floor biases* that activate some neurons regardless of any lack of relevant data points, and the *input layer biases* that accelerate learning on the backpropagation passes.[5]

In the same fashion, the probability that for given neuron states of a hidden layer, a visible neuron is an activated binary *1* with level adjusted by the shared weights w_{ij} modulating the hidden layer neurons h_j is,

$$p(v_i = 1h) = \frac{1}{1 + e^{-(a_i + \sum_j v_i w_{ij})}} = \sigma(a_i + \sum_j h_j w_{ij}) \quad (\text{Eqn.2}).$$

Equation 1 determines the activation probabilities of the hidden neurons (so-called *Gibbs sampling*) for *h* given *v*, where *v* is initialized by a random Bernoulli distribution (binary *yes* or *no*, *1*or *0*, as in a fair-coin toss). Equation 2 determines the activation probabilities of the visual neurons for *v* given *h*; together the equations produce the *joint probabilities* of *h* given *v* and *v* given *h*.[6]

The difference between the initial random Bernoulli probabilities and the input data will be large. Feed-forward runs and iterative backpropagation adjusting the weights w_{ij} and biases b_j will minimize that difference to produce *reconstructions* of the probabilities that will be increasingly better approximations of the input data probability distribution.

In self-supervised learning, the RBM performs forward and backward passes between the visible and hidden layers where the activations of the hidden layer are the inputs to the input layer in a backward pass, multiplied by the same weights, and the sum is added to the input layer bias at each input layer node, and thus constitutes generative iterative reconstructions of the input layer.

The parameters' weighting is best performed not by gradient descent as in artificial neural networks but rather by the so-called *contrastive divergence*. After k iterative runs, the adjusted input values vector v_k is iteratively reconstructed from the original input vector v_o, and used to determine the activation levels of the hidden vectors changing from h_o to h_k. The update matrix ΔW is the difference between the *outer products* $\otimes$ of the vectors v_o and v_k,[7]

$$\Delta \mathbf{W} = v_o \otimes p\left(h_o v_o\right) - v_k \otimes p\left(h_k v_k\right)$$

The new matrix is calculated using *gradient ascent*,

$$W_{new} = W_{old} + \Delta W.$$

For example, if the unknown input data probability distribution $p(x)$ and the reconstructed probability distribution $q(x)$ are both normal distributions but have slightly different shapes and only partially overlap, the difference is the *Kullback–Lieber Divergence* that measures the diverse areas under the two probability distribution curves.[8]

An RBM contrastive divergence minimizes the diverse areas by adjusting the weight and bias parameters to iteratively generate the reconstructed probability, analogous to machine learning.

The integrated difference *Kullback–Lieber Divergence* $D_{KL}(P\|Q)$ of the unknown probability distribution P and the reconstructed probability distribution Q is shown in Figure 24.1.[9]

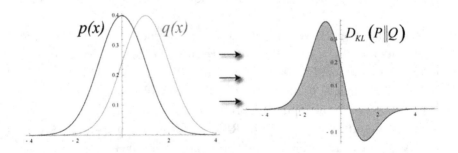

FIGURE 24.1   The Kullback–Lieber divergence.

An RBM is typically used in self-supervised learning initially to first model the unknown input distribution, and then stack the RBMs in a feed-forward network to generatively learn as in a deep belief network (DBN), or act as a preprocessor for convolutional (CNN), recurrent (RNN), or generative adversarial networks (GANs) by substituting its input layer with its hidden layer distribution and feeding its final reconstructed input layer to the input layer of a feed-forward artificial neural network so that the ANN will have a generative-learning head start on determining the input data distribution.

An RBM preprocessor for CNNs, in conjunction with RNNs has been used for speech recognition, but the RBM's main claim to fame is with regard to *latent factors in collaborative filtering* (LFCF, a prime candidate for the most pretentious terminology for a simple concept in artificial intelligence).

LATENT FACTORS IN COLLABORATIVE FILTERING

Linear regression and the restricted Boltzmann machine can reveal the overt features and correlations in input data, but deeper latent inferences may be "hidden" in the data. Stanford University's *Parallel VLSI Architecture Group* has employed machine learning to analyze artificial intelligence neural networks' *hidden layers* to draw "inferences from inference at the data center", meaning obtain hidden information by delving more deeply into not only what the network reveals and infers, but also what *latent* inferences are in the data as inferred by the neuron activation levels h_j in the hidden neuron layer h.[10]

For example, suppose the RBM is presented with data of ages 18–29, users of an online movie website comprising millions of data points gleaned from upvotes (likes), ratings, shares, and peer reviews, *in toto* the film's *buzz*, but the hidden layer feature neurons may also show a cross-correlation revealing a *latent factor* inference for say *redemption* themes, a predilection which the viewers themselves might not be aware.

Netflix classified users into overt *taste clusters*, but based on the hidden redemption inference, Netflix will recommend "you may also like" films that have a redemption theme. Using a restricted Boltzmann machine to inferentially predict the "inferences from inference at the data center"; for example, "twenty movies that are guaranteed to make you cry".

Further processing of cloud-based artificial neural network results can reveal *latent factors* that improve machine recognition that requires a great deal of inference to be accurate, for example in speech, translation, facial expression, body language, and precepts of human behavior.

The complexity of consumer behavior has spawned an entire academic discipline of artificial intelligence marketing psychology, and an enterprising *Adtech* service industry that employs generative learning and Big Data to go beyond marketing studies commonality to delve more deeply into consumer behavior. Adtech's discovery of latent factors driving the deeper psychology of purchasing decisions may result in some very unusual advertisements and commercials.

Some time ago in 2005, Netflix held an open competition for the best artificial intelligence algorithm to predict the attractiveness of new films. The analysis was based on the hard data of film buzz, and the soft data of collaborative filtering; the winning algorithm was performed on a restricted Boltzmann machine.[11]

For natural language processing, an RBM is like a deep belief network (DBN), but acting on probabilities to reveal latent factors in speech from *reconstructed probabilities* that will help it to recognize the implicit meaning of individual speech from context, usage, and all the other vagaries of spoken communication. Then the words and phrases can be collaboratively filtered to include the "hidden" *inferences* so common to natural speech, and thus move closer to more precise NLP.

GENERATIVE ADVERSARIAL NETWORKS

The restricted Boltzmann Distribution is the progenitor of the controversial *generative adversarial network* (GAN). A CNN recognizes an object through *supervised* learning of the training dataset, for example a certain kind of cat. The GAN *generator* network is initialized with a random input that is sampled from an embedding of a set of items within a *latent space* that is a *manifold*, a topological space such that locally is a flat Euclidean space, and in which items resembling each other are positioned closer to one another.

Thus, a *generator* network can *adversely* generate a different image through *self-supervised* learning from the features of the dataset by *deconvolution* ("de-CNN" the reverse of convolution, expands the training dataset for different interpretations of the features). For example, from the features of the dog in Chapter 13, the generator network "creates" in opposition to the CNN's recognition of a cat, a look-alike dog from differently interpreting features overlooked or underemphasized by the CNN.

A *discriminator* in the GAN is supervised-trained on a labeled data set of cats, and attempts to discriminate the generated image from the training dataset image. Both networks are independently backpropagated such that the synthetic (artificially generated) data learned by the generator is spotted by the discriminator, for instance by using a deeper CNN.[12]

A given feature of a dataset may be enhanced by the adversarial generative network, and the image or the like made more accurate by recognizing the possibility of different interpretations. But on the other hand, different interpretations of datasets, text, voice, graphics, and an image's movement and environment are all possible for both productive and nefarious purposes.

For examples, in addition to simulating models on the runway in high-fashion shows and improving the vividness of video games, GANs can perform more esoteric tasks, such as improving astronomy images and simulating gravitational lensing for *dark matter*, which has never been observed, but is believed to constitute 27% of the universe's mass, one of the great cosmological mysteries of our time.

However, GANs can also generate false text, fake voices, graphics, and fake persons (*deep fakes*), all from different interpretations of the same training set data. For example, in 2019, a GAN could generate, from just the contours of someone's clothes, the underlying figure of the wearer. Needless to say, this website was shut down soon after its appearance on the Web.

NOTES

1. RBMs are one of the earliest ANNs and the forerunner of the more recent *generative adversarial networks* (GANs) and generative pre-trained transformer (GPT). The RBM was invented by Geoffrey Hinton, a 2018 Turing Award winners.
2. Methane on Earth is the principal constituent of natural gas.
3. An autoencoder is an ANN that self-supervised encodes input data and decodes it for output.
4. The Π operator (capital π) is just repeated multiplication, just as Σ is for repeated summation.
5. These two biases are conceptually similar to the "hard" and "soft" weights of the Transformer architecture parallel multi-head attention mechanism of generative pre-trained transformer (GPT) described in Chapter 25.
6. Equations 1 and 2 can be derived by applying the Bayes formula (Chapter 30) to the conditional probability equations and expanding. Interestingly, the sigmoid function can be derived from the *unrestricted* Boltzmann

machine from the differences between energy states as expressed by the Boltzmann factor $E_i = -k_B ln P_i$. This shows that artificial neural network probability distributions of objects is mathematically analogous to say the phase changes of matter (solid, liquid, gas) as functions of free energy differences and temperature and pressure.

7. The outer product is a matrix multiplication of two vectors to form a matrix, $[a \otimes b]_{ij} = a_i b_j$.

8. Graph from Mundhenk, T.N. 2009, Ph.D. thesis, University of Southern California.

9. This mathematical description of an RBM has been simplified by the author and Oppermann, A., *Deep learning meets physics, restricted Boltzmann machine, Part I*, towardsdatascience.com, which is based on the original paper, Salakhutdinov, R., A. Mnih, and G. Hinton 2006, *Restricted Boltzmann Machines for Collaborative Filtering*, University of Toronto. RBMs were developed by 2018 Turing Award winner Geoffrey Hinton.

10. Ref. *Spectrum*, IEEE, January 2017.

11. Salakhutdinov, R., A. Mnih, and G. Hinton 2007 *Restricted Boltzmann Machines for Collaborative Filtering*, University of Toronto, Proceedings of the 24th International Conference on Machine Learning dl.acm.org.

12. A deconvolutional neural network runs in reverse to a CNN to regenerate and recover the original input data, increase resolution, reduce signal-to-noise ratio, and test the CNN's ability to accurately extract features from the input data.

CHAPTER 25

LLMs and GPTs

A *Large Language Model* (LLM) is essentially a giant artificial neural network (ANN) that can have trillions of parameters and perform accelerated self-supervised learning from data searched from the Internet using an improved *Common Crawl* search engine.

LLMs takes an input text and predicts the next token, much like natural language processing (NLP), taking an input text and repeatedly predicting the next token or word by various means, such as Hidden Markov Models (HMM) and recurrent neural networks (RNN) and their associated algorithms.

In 2020, OpenAI introduced *Generative Pre-trained Transformer-3* (GPT-3), a large language model that unsupervised scours the Internet in response to a user's textual request (*prompts*), and in addition to semantics, syntax, context, and grammar processing from NLP models and networks, the GPT searches for relevant text from its huge database in response to a prompt, such as "what is a Lagrange multiplier", or "write a poem about birds".

The *generative* output of the model assumes its linear dependence on its own previous values and a stochastic term to form a recurrence relational *autoregressive* model (automatically going back) with *discriminative* (classification demarcation) and NLP fine-tuning to compose prose and poetry, translate, and generate images, music, computer programs, and almost any literal thing that a user requests (now in text and more lately through speech).

Furthermore, the GPT LLM could respond to a user's criticisms about what was produced, such as the writing was too technical or had no logical conclusions, and from *reinforcement learning from human feedback*

DOI: 10.1201/9781003463542-25

(RLHF), the criticism prompts would narrow the domain of the GPT execution and (hopefully) respond with new, more satisfying text for the user.

LLMs take the smallest units of text such as a letter, word, or compound word (*tokens*) and a *byte-encoder* algorithm that forms a series of adjacent letters or phenomes (*n-grams*) such that the most frequent pairs of adjacent characters are merged into a *bi-gram*, thereafter replacing all occurrences of the pair with the bi-gram and thus creates a *relevant* text vocabulary (much like what Pushkin had in mind with his analysis of the letters in *Eugene Onegin*).

Basically, a generative model takes what it has learned and generates some things that are different from the training dataset. The GPT models are neural network-based language prediction models built on the *Transformer architecture.*

The transformer architecture encoder–decoder does not rely on recurrent (RNN) or convolution (CNN) models to generate an output, but instead maps an input sequence to a series of *representations*; that is, assemblages of related information that can be used as a whole to broaden the scope, reduce the search, and better fit the word to a context.

GPTs employ Stanford University's *attention algorithm* that learns the *connection* between *words* and *ideas*, and thus delineates the words in the context of the idea, which offers more accuracy of the words in context, and at the same time better describes the idea.

A transformer requires less training than RNNs, and *long short-term memory* (LSTM), input text is split into n-grams encoded as tokens, which at each artificial neural network (ANN) layer, the token is contextualized as in the CBOW context window by the attention algorithm.

Thus, the slow sequential process of RNNs is speeded up because of parallel processing; that is, the attention algorithm gives all words equal access to any part of a sentence in parallel and therefore can be simultaneously computed.

A transformer architecture is based on mimicking human *cognitive attention* by distinguishing "soft" weights for each word in context (as in Transformers) or sequentially (as in recurrent neural networks). "Soft weights" can change after each epoch, but "hard weights" are pretrained and frozen in what is called the *parallel multi-head attention mechanism*, through which the weighting variability allows the pretraining to come into play to focus on the more salient aspects of the text.

The problems of the vanishing gradient descent and incompatibility with parallel processing of RNNs were solved by the transformer which used *self-attention* to draw global dependencies between input and output by means of the hard/soft weights-based degree of importance and relevancy, and since the self-attention algorithm only uses information on tokens from lower layers, it can parallel process successive layers and quickly provide information about faraway "hard" tokens.

A transformer requires less training than RNNs and *long short-term memory* (LSTM), input text is split into n-grams encoded as contextualized tokens in the CBOW context window by the attention algorithm acting on every ANN layer.

Thus, the slow sequential process of RNNs is speeded up because of parallel processing; that is, the attention algorithm gives all words equal access to any part of a sentence in parallel and thereby has no serial processing latency.

The pretraining of a GPT allows a small textual prompt to generate large amounts of relevant information or to focus on one aspect of information. The GPT models are unsupervised pretrained over unlabeled textual data mainly from the Internet and Cloud by taking some text from the dataset and machine training the model to predict what should come next. The GPTs learn statistical patterns and natural language structure, and can be tuned for specific cases by, for example, the *PyTorch* framework.

GPT-3 has 175 billion neural network parameters with 410 billion byte-pair-encoded tokens, 19 billion tokens from *WebText2*, 12 billion from *Book1*, 55 billion from *Book2*, and 3 billion from *Wikipedia*. In addition to prose and poetry, GPT-3 can program *CSS, JSX, Python* and in principle can code in any language without further training. Later versions such as GPT-4 and its progeny will increase the parameterization numbers even more, new hardware will process more quickly and efficiently, and more LLM data will make the GPT even more *learned*.

New algorithms for GPTs such as generative *diffusion modeling* where a simple starting data distribution employs a sequence of *invertible* operations to simulate a more complicated data distribution.

An *invertible neural network* (INN) can reconstruct the input from the network's output (inverse mapping), as well as the usual construction of the output from the input; that is, it can go both ways, and thus is *bijective*, meaning that two sets are associated such that every member of each set is uniquely paired with a member of the other set (for example, the set of married men to the set of married women is bijective in a monogamous society group).

The INNs are used in probabilistic, generative, and *representation* modeling (which uses a smaller dataset to represent a larger dataset and thus needs less processing and labeling), and may pick out the more significant features of the data through bijection.

The GPT's neural network *ChatGPT* is powered by massively parallel GPUs in an *autoregressive model* (ARM) that performs LLM searches employing *Common Crawl*, and artificial neural networks for natural language processing (NLP).

In 2021, Baidu launched its Chinese language version of *ChatGPT Ernie bot* which has 2.6×10^{11} parameters allowing it to write in traditional and simplified Chinese, solve mathematical equations, answer questions, and even compose classical Chinese poetry.[1]

In late 2022, OpenAI announced the free downloadable ChatGPT (as well as a for-pay premium model) with 1.6×10^{12} parameters and some 800GB of memory; it quickly went viral and was used extensively for all the capabilities mentioned above. But at the same time, unfortunately raised troubling issues of authorship, copyright, discrimination, academic ethics (not really your ideas or work) and so on, and has become a subject of social and academic concern, vigorous debate, and proposed regulation that may hamper its further development.

Virtually all the major tech companies in America, China, Britain, and Europe are continually expanding the databases and upgrading the hardware to thousands of GPUs, and developing new GPU designs for specific uses. For example, Microsoft's *Azure* supercomputer is using 285,000 AMD CPUs and 10,000 Nvidia Tensor Core GPUs, and performs searches at a blazing speed of 316 terabytes/sec.

Users can access a GPT-3 toolset on a text-in/text-out API on *GitHub* from OpenAI, and before long the writers of books and articles, and computer programmers may all go the way of the dodo bird. But …

QUALITY VS. QUANTITY

If one wants ChatGPT to write a poem about birds, can ChatGPT compose Rachel Field's beautiful,

> *Something told the wild geese it was*
> *time to fly*
> *Summer Sun was on their wings*
> *Winter in their cry*

or Coleridge's magnificent *The Eagle* without just copying them?

> *He clasps the crag with crooked hands*
> *close to the Sun in lonely lands*
> *Ringed with the azure world, he stands*
> *The wrinkled sea beneath him crawls*
> *he watches from the Mountain walls*
> *and like a thunderbolt he falls*

Or if one wants pathos plus humor will ChatGPT offer (anonymously),

> *It was a dark and stormy night*
> *A man stood in the street*
> *His aged eyes were full of tears*
> *His boots were full of feet*

That is, scouring the Internet, without simply copying from a database of poems and mixing and matching, can ChatGPT produce *good original* poetry, with message, rhythm, surprise, strong words, and sound. That is, can the GPT machines separate *quality* from just the *quantity* of the database and thorough search in its writing?

"Quality" is difficult to define, but easily discernable, as the above examples illustrate, which will ChatGPT produce?

For image generation, again the Internet is crawled, but this time for images with annotations that match the user's textual request. The simplest is the so-called *zero-shot* text-to-image mapping which finds the image closest to the text description. The *NLP diffusion models (diffusion probabilistic models or score-based generative models)* iteratively employ *latent variable generation* in an attempt to transform overlooked features and noise into relevant data using three steps: the forward process, the reverse process, and sampling, to find latent features that can generate new probability distributions from the training dataset. The learning of the latent structure of a dataset is performed by modeling the way that data points diffuse through their *latent space* (see Chapter 24). It then hopefully will produce higher-quality text, images, and perhaps music, from finding the latent (and therefore deeper or more sophisticated) aspects of a dataset.

Contrastive language-image pretraining (CLIP) is a neural network trained on (image, text) pairs, and given a textual description of an image, it can predict the most relevant text annotation that pairs with the image. The CLIP thereby acts as a bridge between computer vision CNNs and natural language processing (NLPs).

For generating a composite image, beginning with the main subject of the image by for instance using zero-shot, and once found by *Common Crawl*, the GPT purposefully overlays random visual noise onto each subject in the image, and then uses regular machine learning to excise the noise, and the generative network thus has learned "denoising" for each subject in the composite image, as shown in Figure 25.1.[2]

The user then will find another subject in the composite image, and repeat the process of noising and denoising, with GPT learning how to excise the noise of that part of the image, and so on, "learning" to produce clear images of each subject. Then after a user prompt, the GPT will recall all the images at once, and since it knows how to remove the noise from each of them, will (hopefully) present the composite image desired.

Generating music employs a *MIDI* file generation model and *latent diffusion modeling* to match prompt requests, finding relevant snippets from the music files, and combining them. Of course, GPT will have problems sequencing notes and movements in a sonorous and appealing combination.

A *latent diffusion model* generates music similarly to the generation of composite audio, but uses a *variational autoencoder* to map audio onto an *embedding* (generally, representing data points in a lower-dimensional space) of 800,000 audio files including music, sound effects, and instruments. The model then will process and stream back the generated music with annotations. However, it will not likely ever replace original, good music by talented composers.[3]

The ability to talk to ChatGPT draws on two separate models. *Whisper*, OpenAI's existing speech-to-text model, converts what you say into text, which is then fed to the chatbot. A new text-to-speech model converts

FIGURE 25.1   Cat image noising and denoising.

ChatGPT's responses into spoken words. However, although more accurate, GPTs will have difficulty avoiding all the problems of NLP in robot-to-human, and text-to-speech recognition.

Start-up Rabbit's voice-controlled prototype called "*n*" allows users to tell or gesture commands, and learns from the user's previous actions so that whereas formerly users had to install and learn how to operate apps, now AI learns from the user, to provide services.

GPT-4 has one trillion parameters (six times GPT-3) and uses thousands of GPUs to scour hundreds of billions of words for weeks, and since it requires massively parallel matrix computations by GPUs, the transistors must do huge amounts of switching at once which requires much more electricity than serial processing. It has been estimated by OpenAI that training GPT-4 costs 50 gigawatts of electricity (about 50 times more than GPT-3 and equivalent to about 0.02% of the total amount of electricity generated by California in a year). Furthermore, the fast parallel processing means great heat is generated and the water-cooled servers require vast amounts of water. Altogether, GPT-4 training is estimated to cost about $6m a shot.

From the point of view of economics then, researchers are trying to get more performance out of fewer GPUs and smaller databases. There have been projects to round-off numbers, focus on specific tasks, freeze the first run of parametrization, and insert smaller models in-between to generate a large, general "teacher model" to teach smaller, more specific task models, reduce dataset sizes by *directed data mining*, and utilizing existing *distributed algorithms* for similar tasks.

AI computer coding can more efficiently use the GPUs, and even redesign or replace the GPUs with those specifically designed to run LLMs and GPTs, for example, Google's tensor processing unit TPU could be modified for more efficient processing of LLM GPTs, and Advanced Micro Devices' (AMD) M1300 chip has 153 billion transistors, 192 GB memory, and 5.3 terabytes/sec bandwidth, all about twice the specs of Nvidia's highest-end GPU.

Software also could be improved to use the GPUs more efficiently, for example, Google's Cloud TPU platform and Meta's *PyTorch* framework could provide more details on how computations are arranged on the chips. GPUs could be designed for efficient GPT processing, for instance, Google's TPU, Meta's *MTIAS*, and Amazon's *Inferentia* chips are more suitable for massively parallel GPU processing.

GPTS AND SCIENCE

In 1665, Robert Hooke saw the then new microscopes and telescopes of the times as "adding artificial organs to the normal". Scientists could see much smaller details, and very farther things with the new instruments, and in doing so, understand more about the Earth's natural world and indeed the universe.

This "extended capability" as developed in artificial intelligence already has been realized in the many machines described in previous chapters, and more recently to deconstruct the molecular structure involved in protein-folding to discover and design new drugs (for instance, to mitigate the scourge of viruses such as Covid-19), new materials for batteries, control of nuclear fusion reactions, and prospecting for hydrogen in the Earth's crust, perhaps leading to truly clean and abundant electricity sources, and so on.[4]

The computer-based technology, heretofore, has been largely limited to scientists and engineers who know how to write computer code, but with the advent of large language models and GPTs, an AI machine such as ChatGPT is now a tool that virtually anybody can source a program by simply describing what they want it to do.

For instance, the time-consuming and tedious survey of the scientific literature to find correlations and perhaps new discoveries for further research can be handled by GPTs 24/7, hopefully leading to more edifying work for graduate students, or a plethora of new robot scientists, like the robot Darwin described earlier.

ChatGPT will even have a *robots for romance* that is not a dating service, but advice on how to build a relationship for couples that will be mutually beneficial, satisfying, and enduring.

The question however remains, does knowing a lot, and the ability to match and process information, constitute *intelligence*, and if not, what does?

NOTES

1. A particular problem with voice input is the very large number of homonyms in the national Mandarin language, which may only be solved by more contextual processing.
2. Cat image from Stanford University, Computer Science, CS 231n_02 Spring, personal communication.
3. For examples, for a symphony, *allegro, adagio, scherzo,* and *sonata-allegro,* and for popular music a clear and emotive theme and driving rhythm.
4. Ref. *The Economist* 6/24/23 and 9/16/23.

CHAPTER 26

Massive Parallel Processing and Supercomputers

THE VON NEUMANN COMPUTER architecture is serial; a Central Processing Unit (CPU) processes the programs and data that are fetched from a memory storage unit through a bus connected to the central processing unit (CPU). Each piece of information is assigned a memory location with a unique address, and after being fetched, is processed sequentially by instruction cycles in step with a timing clock. Because all fetched information shares the bus, the CPU must always wait for instructions and data before it can proceed, which can slow down operations.

Many speedup techniques for the von Neuman architecture have been employed to decrease the *latency* (time between procedures), including adding an input/output processor, partitioning memory into banks, installing fast data caches, adding a coprocessor to perform some slower functions faster, pipelining for multiplex operation, and multiple CPU cores, as shown schematically in Figure 26.1.[1]

With these additions, von Neumann machines can handle most of today's routine computing tasks using multicore serial architecture where, of the LN's designated layers, LLC is the last layer cache and is shared among the cores, and if the data is not in the caches, it will be fetched from the DDR-4 memory.

DOI: 10.1201/9781003463542-26

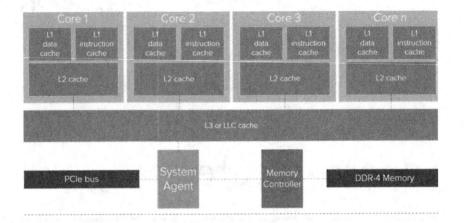

FIGURE 26.1   Multicore serial processing computer architecture.

The multiple cores naturally led to the idea of a tensor architecture where the processor treats information as a vector or matrix of data elements instead of individual scalar data points, and in place of a single or multicore processor, an *array* of processors can simultaneously proceed in parallel, and need not wait in line for the system bus to arrive to begin processing, thereby greatly decreasing latency, as shown in the flow chart of Figure 26.2.

GPUs were originally designed for the dynamic images of computer games, where an array of pixels (a *raster*) defines the images, and animation of the images is the change in the image attributes (including textures and shading) with time. *Ray-tracing* generates an image by tracing the path of light through the pixels in an image plane producing realistic 2D renditions of 3D objects by using different ray paths; the pixels are stored in a matrix and GPUs comprising multiple core CPUs perform the calculations in parallel simultaneously.

The fast movements of the agent and the NPCs are easily blurred, and ray-tracing, rasterization, and vector graphics more clearly focus the image as shown in Figure 26.3.[2]

The parallel-processing architecture is schematically shown in Figure 26.4. Each processor cluster bus (*PCle*) is connected to multiple streaming multiprocessors (*SM*) and each SM has a layer-1 instruction cache layer. One SM will fetch from a dedicated layer-1 cache and a shared layer-2 cache before using the global GDDR-5 memory; the cache layers in GPUs are generally fewer and smaller as GPUs are less concerned with memory latency as long as the parallel processing is going on smoothly; from this,

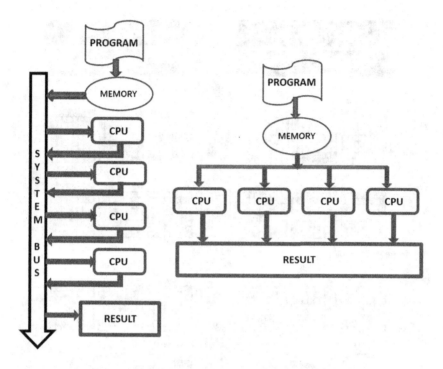

FIGURE 26.2   Schematic serial and parallel-processing flowchart.

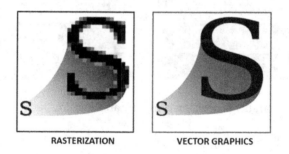

RASTERIZATION   VECTOR GRAPHICS

FIGURE 26.3   Rasterization and vector graphics.

the GPU parallel-processing architecture can greatly increase computer throughput.[3]

A typical deep artificial neural network can process 30 GBytes of data using millions of nodes. Utilizing GPUs instead of CPUs can reduce the number of nodes by two orders of magnitude. But some processing requires knowing the result of one step in order to process the next step and thus are inherently serial, for example, conventional cryptography algorithms and other iterative operations cannot be parallel-processed.

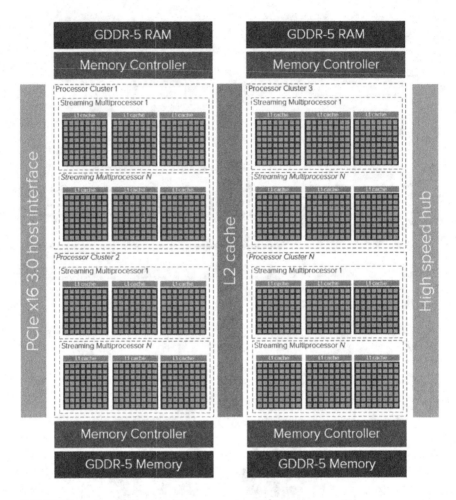

FIGURE 26.4   Parallel-processing computer architecture.

In parallel-processing, the *average parallelism* is defined as the *Work* over *Depth*, which is just the total number of operations, divided by the Depth of the network (the total number of layers in the network),

$$Average\ Parallelism = Work/Depth.$$

Because of the huge amounts of data, deep learning data processing requires a prodigious number of operations, so *mini-batches* are processed in turn. Taking mini-batches will also promote greater accuracy because noise will be filtered out on average over the mini-batches. Finding the right size for the mini-batches is an exercise in just seeing what works best for the task at hand.[4]

The limit to how much a computation can be speeded up by parallel-processing is given by *Amdahl's Law,*

$$Maximum\ Speed\ Up = S = \frac{1}{1 - P + \dfrac{P}{N}}$$

where P is the proportion of the system that can be made parallel (and $1 - P$ is the proportion of the system that remains serial) and N is the number of processors. As the number of processors $N \to \infty$, $S \to 1/(1 - P)$, so the *SpeedUp* ultimately depends on the proportion of the system that cannot be parallel-processed; this can never be zero as it is unavoidable that some operations depend on the results of computations that came before.

The deep convolutional neural network *GoogLeNet* that won the 2014 ImageNet Large-Scale Visual Recognition Challenge had 6.8 million parameters and is 22 layers deep, running on the Nvidia Tesla V100 with 80 Streaming Microprocessors each with 64 cores. Software to run the Teslas included *VMWare VSphere ESXI* that dedicates a GPU to a Virtual Machine using *DirectPath I/O.*

Google's *tensor processing unit* (TPU) is architecturally the same as a GPU but does not perform the graphical rasterization or texturization; it is specifically designed for machine learning neural network operations using the *TensorFlow* platform, mundanely in high volume, low precision processing such as *GooglePhotos*, and sensationally as in AlphaGo's defeating the legendary Lee Sedol in a world championship *Go* match.

The TPU was specifically designed for machine learning neural network computations using the *TensorFlow* platform. It has progressed from a simple 8-bit matrix multiplier chip fabricated in a common 28 nm process, to adding 600 GB memory bandwidth with performance each reaching 45 *teraFLOPS* arranged in four-chip modules, and doubling the processing power using 16 chips per module.

In TensorFlow's *data graphs*, every node is a tensor (which is higher than a rank 1 (number of indices) vector representation), and following the algorithm, it can rapidly parallel process data, probabilities, and functions to find correlations and the influence of interrelationships.[5]

Scaling deep machine learning requires programming clusters of processors, a complicated exercise at best; a single chip allows simpler programming and faster computation. The world's biggest and fastest single AI chip is the second-generation Cerebras *Wafer Scale Engine* (WSE-2) which is 56 times larger, has 3,000 times more memory, and 10,000 times the memory bandwidth of Nvidia's largest GPU.

WSE-2 has 10^{12} transistors, 850,000 cores, 40 GB on-chip memory one clock-cycle away from the cores, and 100 petabits/second memory bandwidth (rate at which data read from and stored to memory) comprising a low-latency, high-bandwidth input/feedback chip specifically designed for machine learning, and purportedly will reduce the interconnection transmissions latency from weeks to minutes.[6]

While the capabilities of even super-thin notebook computers have already far surpassed the bulky early acronymic mainframes, much greater computer speed and power are required for the massively parallel processing of the trillion-parameter, LLM GPTs, and the differential equations of scientific artificial intelligence.

SUPERCOMPUTERS

The idea of a supercomputer began in the 1970s, with Control Data Corporation and Cray Research vying for the title of world's fastest computer. The CDC and Cray met competition first from Japan's NEC and Fujitsu, then later from China's Sunway and Tianhe, taking turns with America's IBM Summit and Sierra for the speed crown. The race has become symbolic of national technological prowess.

Supercomputers have been used for Hydrogen Bomb simulations, weather forecasting, molecular structure analysis, stellar evolution, fluid dynamics, genetics, cryptanalysis, cosmological calculations of the origins of the Universe, and now critically for climate change, global warming, protein-folding, and indeed, as Microsoft's current use of its *Azure* supercomputer for OpenAI's GPT-4, for processing the Big Data/LLM/GPT computations of artificial intelligence.

A supercomputer typically simultaneously solves a dense $n \times n$ system of linear equations $Ax = b$ often encountered in science and engineering. The current (2024) fastest supercomputer is IBM's Frontier supercomputer that achieved a R_{max} (largest problem in GFLOPS) of 1.102×10^{15} floating point operations (exaFLOPS), breaking the exascale benchmark) (Figure 26.5). Frontier uses 9.472 64 core AMD CPUs and 37,888 Radeon Instinct GPUs having 8,335,360 cores.[7]

For an example of speed, solving the equations of stellar evolution for a star from birth to death would take an astrophysicist 3,000 years, while a supercomputer can do it in seconds. Closer to home, simulations of nuclear weapons explosions could allow nuclear-armed belligerents to

FIGURE 26.5   IBM's frontier supercomputer.

just exchange simulation print-outs and computer graphics to determine whose bomb was more destructive, bringing warfare from the battlefield to the laboratory, if only world leaders knew something about science.[8]

NOTES

1. Serial and parallel architecture figures and descriptions use kindly provided by Hagoort, N., "Exploring the GPU architecture", nielshagoort.com.
2. Rasterization and vector graphics image from an anonymous artist in the public domain under Wikimedia Commons.
3. GDDR-5 is a "graphics double data rate type 5 synchronous random-access memory".
4. Hoefler T., 2019, *Demystifying Parallel and Distributed Deep Learning*, Swiss High Performance Computing (HPC) Conference.
5. A *tensor* includes rank 0 (scalars), rank 1 (vectors), rank 2 (matrices); and higher rank tensors depend on the number of indices required to select each element, for example Einstein's gravitational field equation is described by the four indices Riemann curvature tensor. See Chen, R.H. 2017, *Einstein's Relativity, the Special and General Theories with their Cosmology*, McGraw-Hill Education (Asia).

6. The "Wafer" indicates that instead of the size of semiconductor chips cut from a pure silicon wafer, the WSE is the size of the wafer itself.

7. Specs for supercomputers vary according to how architecture configured and speed is measured; the peak speed is theoretical and operational speeds are typically 30–50 petaflops slower. Frontier image by OLCF at ORNL – https://www.flickr.com/photos/olcf/52117623843/, CC (Creative Commons) BY 2.0, https://commons.wikimedia.org/w/index.php?curid=119231238. High Performance Linpack (HPL) measures supercomputer solution speeds. A *petaflop* = 10^{15} floating point operations per second. Although hard to grasp, such fantastic speeds are possible because of the very short atomic distances (about 10^{-10} m) in semiconductors, so the electrons and holes don't have far to travel to do their work, which they can perform at about half the speed of light.

8. The author has programmed Cray supercomputers and suggested the print-out exchange in lieu of Hydrogen Bomb Armageddon, receiving no response.

CHAPTER 27

Quantum Computing

Q UANTUM COMPUTING CAN PROCESS extremely complex and tortu- ous equations and combinatorial problems at lightning speed, based on the formation of quantum mechanical wavefunctions, their superposition, coherence, decoherence, and collapse, all in abeyance to the basic principles of quantum mechanics relating to probability amplitudes and the conveyance of information.

WAVE/PARTICLE DUALITY

Because of its straight-line propagation and reflection, Newton in the 17th century believed light to be *corpuscular*, but Thomas Young's 1802 double-slit experiment definitively demonstrated the *wave* nature of light.

Light of the same wavelength (color) passing through two slits in a screen produced interference fringes on a detector screen, revealing the constructive and destructive interference characteristic of phase difference superpositions of the timing and interval difference of waves traveling through the slits, as shown in Figure 27.1.

Furthermore, in the 19th century Maxwell mathematically proved that light is an *electromagnetic wave* derived from his four famous electromagnetic equations.

But coming full circle in 1900, Planck's small bundles of energy (quanta) and Einstein's photoelectric effect in 1905 proved that light was after all corpuscular, to be called a photon, just as Newton believed.

In 1924, however, de Broglie using streams of electrons reflected by crystals and diffracted by thin metal foils, astoundingly showed that a *particle*, the electron, with momentum (p) and energy (E), just like light

DOI: 10.1201/9781003463542-27

 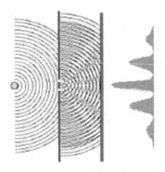

FIGURE 27.1   Young's double-slit experiment showing the wave nature of light.

is diffracted by double slits, and produces interference fringes just as in Young's double-slit experiment for light; that is, an electron is both a particle and a *wave*, with wavelength (λ) and frequency (f), related through Planck's constant h,

$$\lambda = h / p \text{ and } f = E / h.$$

The wave–particle duality has long been one of the mysteries of physics that an object or phenomenon can be described as either a wave or a particle, but not at the same time or in the same situation.

One of the founders of quantum mechanics, Werner Heisenberg, explained the dissonance arising from this duality that something can be one thing and entirely another thing depending on how you look at it,[1]

> *[Analogies] may be justifiably used to describe things for which our language has no words … It is not surprising that our language should be incapable of describing the processes occurring within the atoms, for … it was invented to describe the experiences of daily life, and these consist only of processes involving exceedingly large numbers of atoms. Furthermore, it is very difficult to modify our language so that it will be able to describe these atomic processes, for words can only describe things of which we can form mental pictures, and this ability, too, is a result of daily experience.*
>
> *Fortunately, mathematics is not subject to this limitation – the quantum theory – which seems entirely adequate for the treatment of atomic processes; for visualization, however, we must content ourselves with two incomplete analogies – the wave picture and the corpuscular picture.*

In experiments, one electron (or photon) at a time projected from a source through the double slits to an interferometer showed the characteristic interference fringes of waves and therefore the electron is in a *superposition* of two states, one state of passing through slit A and another state of passing through slit B, exactly which slit is passed through cannot be determined because there is only a *probability* of passage through each of the slits and there are myriad possible combinations of the different probabilities.[2]

If an electron, or other object, can demonstrate superposition, can a *cat* be in a superposition between two states as well?

THE SCHRÖDINGER WAVE EQUATION AND DIRAC BRACKETS

The quantum mechanical behavior of particles, such as the photon or electron (disregarding spin) is governed by the *time-dependent Schrödinger equation* for quantum mechanical *wavefunctions* Ψ which present the probabilities of the particle's position, momentum, and time-dependence, from their energy and time,

$$-i\hbar \frac{d}{dt}\left|\Psi(t)\right\rangle = \hat{H}\left|\Psi(t)\right\rangle,$$

where $\hat{H}$ is the *Hermitian (self-adjoint) Hamiltonian* of the total energy of the system and $\hbar$ is the so-called *reduced Planck constant* $h/2\pi$. Because the wavefunction describes the probability amplitude of an event, the Schrödinger wavefunction is not a physical wave like water. Only *standing waves* that fit wholly within the period of the wavefunction are allowed. Thus only solutions having certain wavelengths and integer multiples of those wavelengths (harmonics in music) are allowed; they are thus *quantized* as stable *eigenstates* of the Schrödinger wave equation, wherein the associated *eigenvalues* of the equation are quantized, real, measurable quantities, called *expectation values*. The amplitudes of the eigenstates' complex conjugates squared represent the probability of the occurrence of the event.[3]

Another formulation of quantum mechanics was devised by Heisenberg and Max Born using matrix operators operating on state vectors. Those vectors are conveniently expressed in the *Dirac notation* using the *ket* $\left|\Psi\right.$ state vector which represents a state in a quantum system, and when projected onto a corresponding *linear form dual bra* $\left\langle\Phi\right|$ is denoted by $\left\langle\Phi|\Psi\right\rangle$, which is a complex number that determines the linear dependence of the dual and the state vector $\langle \Phi | \Psi \rangle$.

From this, predictions can be made as to the characteristic of the behavior of the system. Simple examples are the projection of the state vector $|\Psi\rangle$ onto its complex conjugate $\langle \Psi |$, which produces a real number, and the projection onto a *covector* which takes a vector and turns it into a scalar; that is, is a linear projection from vectors to scalars, for instance the inner product.

The Dirac notation convenience is that any state (for instance, photon polarization, electron position, quark momentum, particle spin, and so on) can be simply inserted into the ket $|\Psi\rangle$ state vector and then linear correspondences between the state vectors and *forms* can be simply expressed by $\langle \Phi | \Psi \rangle$.[4]

For an example in the Heisenberg/Born formulation, a matrix operator M representing some phenomenon, operating on a state vector is expressed as $M|\Psi\rangle$, and the eigenvalue equation can be simply written as

$$M|\Psi\rangle = \lambda|\Psi\rangle,$$

where λ is the energy eigenvalue. If the matrix operator M is *Hermitian*, then the eigenvalues will be real; meaning that Hermitian matrix operators bring reality into a wave equation by relating it to a measurable observable.[5]

A state vector $|\Psi\rangle$ can be represented generally by an expansion in a linear combination of an orthogonal basis vector set $|j\rangle$, constituting a *superposition* of an infinite number of possible states, which are the eigenstates of a complete ensemble of commuting observables.

$$|\Psi\rangle = \sum_j |j\rangle\langle j|\Psi\rangle,$$

A measurement of the complete ensemble randomly projects the eigenvector onto one of the j states with a probability of $|\langle j|\Psi\rangle|^2$, and so $\langle j|\Psi\rangle$ is a *probability amplitude* for that state. Since it is a probability, $\langle j|\Psi\rangle$ must be normalized to a closed interval [0,1] where the lowest number 0 means that the event will not happen, and the highest number 1, means the event will absolutely happen.

These probability amplitudes interfere like electromagnetic waves in classical physics, however, *it is the probability amplitudes* of the particles' quantum mechanical wavefunctions that *interfere* (producing superpositions of the wavefunction), and not the particles themselves. That is why it is not possible to identify a particular particle passing through a given slit; the behavior is only probabilistic and not deterministic.

A quantum mechanical wavefunction must be continuous, so the summation becomes an integral over all the possible continuous states in that basis set,

$$|\Psi\rangle = \int |j\rangle\langle j|\Psi\rangle d^3 j,$$

The shape and extent of the wavefunction depends on *close* interaction with the environment (usually a measurement). An electron's wavefunction has dimensions of a Bohr radius 5.3×10^{-11} meters, the average radius of an electron's orbit around a hydrogen atom in its ground state; this is a close interaction distance.

Different wavefunctions originating from the same event are in superposition, the relational properties between the multiple waves are *coherent* (like a laser) when they are in-phase or have a constant relative phase plus the same frequency, and can interfere with each other; loss of any of these relationships causes the wavefunctions to *decohere*, which eventually will happen due to the impact of the environment, and the wavefunction information will be lost.

Upon a measurement of say position x, the wavefunction is given by $|\Psi\rangle$ mapping on to scalar position,

$$\Psi(x) = \langle x|\Psi\rangle.$$

When placed inside an integral over all space, $\langle x|\Psi\rangle$ is the *probability density* for wavefunction *collapse* to x, a real value,

$$\int |x\rangle\langle x|\Psi\rangle d^3 x.$$

If the measurement is performed with a precision of δx, the superposition of states immediately collapses with a probability of $|\langle x|\Psi(x)\rangle|^2 (\delta x)^3$ into a small wave packet of volume $(\delta x)^3$ around the position x. Thus, when a measurement is performed, an arbitrarily extended wavefunction is collapsed into a wave packet with a much reduced extent; that is, a system by *closely* interacting with its environment is transformed from a superposition of states into a definite state with a well-defined value of the measurement.

For a momentum measurement, the wavefunction is similarly expanded over a linear combination of the basis set of momentum eigenstates $|p\rangle$. And of course, anything having to do with waves will employ a Fourier Transform; that is, each momentum eigenstate can be represented as a superposition of its conjugate *position* states, and the de Broglie particle wavelength $\lambda = h/p$ as a function of momentum p.

The momentum wave function thereby can be represented by a plane wave as a Fourier series over position x, and since the quantum wavefunctions must fit as standing waves and are continuous, they can be described by a *Quantum Fourier Transform* algorithm (QFT),

$$|p\rangle = \int |x\rangle\langle x|p\rangle d^3x = \frac{1}{(2\pi\hbar)^{3/2}} \int e^{ip\cdot x/\hbar} |x\rangle d^3x,$$

where the factor before the integral is the quantum Fourier transform coefficient. The attributes of Fourier Transforms such as the speed-up Fast Fourier Transform (FFT) thus may be gainfully used for computations of quantum mechanical wavefunctions.[6]

The momentum in terms of its position is just a change of basis vector set,

$$|\Psi\rangle = \int |p\rangle\langle p|\Psi\rangle d^3p,$$

then the complex conjugate is

$$\tilde{\Psi}(p) = \langle p|\Psi\rangle = \int \langle p|x\rangle\langle x|\Psi\rangle d^3x = \frac{1}{h^{3/2}} \int e^{-ip\cdot x/\hbar} \Psi(x) d^3x.$$

Just like the position, a momentum measurement has probability density $\left|\langle p|\tilde{\Psi}(p)\rangle\right|^2$, and from the measurement, the wavefunction of the particle collapses into a momentum eigenstate represented by a plane wave wavefunction, and the state of the wavefunction can then be observed.

SCHRÖDINGER'S CAT

The *Copenhagen school* of quantum mechanics led by Niels Bohr believed that before an observation is made, the state of the wavefunction, being a superposition of states, cannot be known until collapsed by an observation, and therefore there is no way to determine any objective reality before then.

Schrödinger's cat was in a box with a radioactive atom, which has a 50% probability of beta-decay within one hour. Upon emission of an electron (beta-decay), a circuit will be closed causing a hammer to fall and crack open a vial of poison gas. When and if an electron is emitted during that hour cannot be known because it is quantum-mechanically probabilistic. So during that hour, Schrödinger's cat is neither dead nor alive, but is in a superposition of the two states of life and death (Figure 27.2).[7]

This appears to be at odds with common sense, so no wonder Einstein fought long and hard against the Copenhagen school, as he said, "physics should represent an objective reality in time and space", and criticized the *Copenhagen school* saying that "there is an objective reality of the state of the cat in the box during that hour, even if we did not observe it".[8]

But the Copenhagen school saw the cat as a single wavefunction which has a superposition of the two states, alive and dead, just like the single photon having a probability of passing through each slit, and is not definitively determinable until the wavefunction collapses by a close interaction, an *observation*,

$$\Psi = \left| livecat + deadcat \right\rangle$$

This makes quantum mechanical sense, but is regarding the macroscopic cat as a microscopic entity, and only the macroscopic cat can be observed in this case. Schrödinger's cat actually is just a macroscopic rendering of microscopic quantum mechanical superposition, but it led Einstein (objecting to the analogy) to say that,

I cannot believe that God is playing dice with the Universe.

or even worse, are the dice-playing God?

SCHRÖDINGER'S CAT IS
A⃒L⃒E⃒⃒A⃒V⃒⃒E

FIGURE 27.2   Schrödinger's cat in a state of superposition.

In sum, an explanation perhaps of comfort to Einstein (as a cat owner) is that beta-decay is a *probabilistic* microscopic phenomenon, but a cat's life or death is a *deterministic* macroscopic phenomenon with an *objective reality*. That is, Schrödinger mixes the microscopic probabilistic nature of quantum physics, and the objective macroscopic state of the cat of classical physics, so the cat's fate is an objective reality and hopefully is alive and well.

The micro/macro mixed phenomena cannot be represented by a quantum mechanical wavefunction alone, but if a macroscopic *density operator* is employed, then a classical statistical probability can be used in an experiment with sufficiently large sample size, say 1,000 cats in 1,000 boxes, will produce a predictive analytic statistical result for the sample, but just like an average, never a definite result for a specific cat.

Although both are based on probability, the quantum mechanical treatment is for one event, while predictive analytics requires the Law of Large Numbers to determine probabilities, but like quantum mechanics cannot predict the occurrence of a single given event.

To animal lovers, it is important to realize that Schrödinger's cat of course is just a metaphoric thought experiment, but the macroscopic statistical experiment is rightly unlawful and liable to justifiable condemnation by the society for the prevention of cruelty to animals (SPCA).

QUANTUM ENTANGLEMENT

The *probability amplitude* of one state vector $|\Psi\rangle$ projecting onto a *form dual* $\langle\Phi|$ through a matrix operator M for some phenomenon is given by $\langle\Phi|M|\Psi\rangle$, and the actual *probability* of that *event* happening is given by the amplitude and its complex conjugate squared to produce a real number,

$$\text{Probability} = \langle\Phi|M|\Psi\rangle \cdot \langle\Phi|M|\Psi\rangle^* \equiv \left|\langle\Phi|M|\Psi\rangle\right|^2.$$

When the probability as expressed above is multiplied out, there may be *cross-terms* which express the probabilities of *Verschränkung* (correlations) within the *wavefunctions*; these cross-terms give rise to the mysterious *quantum entanglement*.

A simple analogy is a pair of gloves are in a box; you take out the left glove and send the box to an astronaut on Mars, and as soon as the box is opened, the astronaut instantaneously knows that you have a right glove.

This transmission of information is faster than the speed of light, and according to Einstein's Special Theory of Relativity, faster than light conveyance of information is impossible; even more uncomfortable with this than the dice-playing God, Einstein called it *spukhafte Fernwirkung* ("spooky action at a distance").

The crux is that the two states were *entangled* to begin with because gloves always come in pairs; so the right-hand glove sent to Mars instantly conveys the information that the sender has a left-hand glove.

The wordplay solution is that information has not really been *transmitted* but *revealed*. A more technical example is when two photons are emitted as the result of one single event, for example, a crystal which absorbs a photon of twice-resonant frequency and instantly emits two photons, one red and one blue, with energies adding up to the energy of the absorption. The photons upon emission are neither red nor blue, but a superposition of red and blue (purple?), and because of their mutual origin from a single crystal absorption and emission event, they are *entangled*.

If the red photon wavefunction is collapsed by an observation, then the wavefunction of the other photon also immediately collapses and will be a blue photon, because the red and blue photons were originally entangled, and the information that the other photon is blue is instantly transmitted.

The first experiment that verified entanglement was performed in 1949 by Chien-Shiung Wu and I. Shaknov on the correlation of angular momentum entanglement of two photons traveling in opposite directions. Other experiments on photons, electrons, and even small diamonds have definitively demonstrated quantum entanglement.

Experiments agree that entanglement produces a correlation called an *emergent property*, and that the mutual information between the entangled particles can be exploited, for example, for exponentially increasing calculation, combinatorial explosions, satellite communication, cryptology, and even teleportation (as in science fiction's bodily decomposition and re-composition at another time and place).

Life and death, right and left, red and blue are binary like 0 and 1, but between the binary states are an almost infinite number of coherent superposition states (depending on the number of qubits); and that information may be read from the wavefunction collapse and quantum entanglement is the crux of quantum information systems and quantum computing.

INFORMATION PROCESSING BY QUANTUM COMPUTERS

Unlike classical computers binary bit 1 and 0, quantum computing uses the *qubit* which is the basic unit of information in quantum computing.

For example, a two-qubit system has four basis states as a two-state *unitary matrix* (complex vector of size 2), where $|\psi\rangle = \alpha|0\rangle + \beta|1\rangle$, and α and β represent the probabilities of the qubit being in states *0* and *1* with the probability normalized to *1* by $|\alpha|^2 + |\beta|^2 = 1$. Qubits in quantum logic gate arrays are used for input/output and computation.[9]

A physical qubit is a superconducting tiny metal loop or wire made of aluminum, or an alloy of aluminum and niobium that behaves like an atom. It is created by laser or microwave irradiation, or subjected to strong magnetic fields, producing *mixed state* wavefunctions that exhibit superposition, interference, entanglement, coherence, and decoherence, all of which are used in quantum computing.

A simple analogy of a mixed state superposition is a coin flipped in the air is in a mixed state of rotating heads and tails, and only when hitting the ground, stops rotating and either head or tails can be observed (the ground *collapsed* the "coin wave function"), and at the same time, the other side is also known because a coin has heads and tails which information is entangled in the coin structure itself.

For *n* qubits, there are 2^n possible superposition states, for example for $n = 50$, $2^{50} = 10^{15}$ superposition states, and before observation, the superposition states and their combinations are entirely unknown (from this, a cryptological quantum computed message is impossible to surreptitiously decode).

Quantum computers operate using qubits, and linear algebra of matrices rather than the Boolean Algebra of classical computers. Unlike traditional bits which can only be *0* or *1*, a qubit can exist in a mixed superposition of *0* and *1* with a huge number of mixed states, instead of only the *1* and *0* of classical computers.

The *Bloch Sphere* displays the qubit states. The north pole in Dirac notation is the state vector $|0\rangle$ and the south pole $|1\rangle$ are pure states *1* and *0*, and all the other states inside the sphere are mixed states. The points on the equator have equal probability of being north (*0*) or south (*1*) pure states, the point in the middle of the sphere indicates a completely mixed state. The position of the point in the Bloch ball indicates the relative probabilities of the qubit yielding a pure state *1* or *0* (Figure 27.3).

That the superpositions of the qubit all arise from the qubit itself, the mixed states are entangled, and when the wavefunction collapses, the entangled emergent properties can instantly manifest themselves even if a great distance apart.[10]

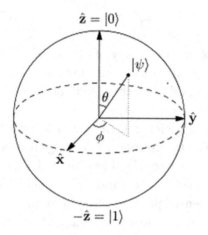

FIGURE 27.3   The Bloch Sphere (ball).

An explanation suggests that there are "hidden variables" in the quantum wavefunction revealing the emergent property information, but this evokes a more troubling implication that the theory of quantum mechanics is "incomplete", debated and lamented by among others, Einstein, presenting a controversial issue in quantum mechanics.[11]

But all agree that ultra high-speed quantum computers can be used for simulations of high-complexity combinatorial systems such as the very large number of chemical compounds and interactions in computational chemistry, materials science, and drug design, complex atomic and molecular physics systems, cryptology, encoded satellite communications, even eventually in *teleportation*, first of objects, and then possibly of humans.

A qubit artificial network based on qubit perceptrons, together with massive parallel processing, huge memory capacity, and high-complexity quantum logic gate matrices cascades, can perform machine learning on a set of continuous parameters comprising the superposition states between *1* and *0*, and of course, the more states and parameters, the more accurate the simulations, and the utility of the machine itself.

However, the immense number of gates and combinatorial complexity in the electronics and the environment will produce *noise*, resulting in changes of a wavefunction's quantum states (*decoherence*) and consequent loss of information. Therefore, to prolong coherence, the quantum computer requires supercooling and all the other special equipment needed for superconductivity, plus software to reduce noise and increase fault tolerance.

As early as 2008, the University of Science and Technology of China devised a *quantum relay* which could transfer a decaying photon signal to another photon, and thus set the record for the longest transmission of quantum computing information.

In 2016, the China National Space Center launched the *Mozi* satellite, the first to use qubit superposition and entanglement to encode information from a satellite which could not be read until it reached its destination.

In 2020, the Hefei Science and Technology University *Jiuzhang* qubit information tensor processor took only three minutes to process *Gaussian Bose Sampling*, a feat that would have taken at that time the fastest supercomputer in the world, the *Sunway TaihuLight* a billion years to complete.[12]

In 2021, IBM announced its $n = 422$ qubit superconducting quantum computer *Osprey* with 2^{422} possible superposition states, an unimaginably high exponential number, but IBM in 2023 expects that its new *Condor* quantum computer will reach $n = 1,121$ qubits, and researchers are expecting even a hundred thousand qubit machine, in a bid to win the race for quantum computing supremacy, which although being a marvel of technology, can appear menacing as well (Figure 27.4).[13]

FIGURE 27.4   IBM Osprey $n = 422$ quantum computer.

NOTES

1. For Maxwell's equations and light, see Chen, R.H. 2011, *Liquid Crystal Displays, Fundamental Physics & Technology*, Wiley; for quantum mechanics, see Chen, R.H. 2017, *Einstein's Relativity, the Special and General Theories with their Cosmology*, McGraw-Hill Education (Asia). Quote from Heisenberg, W., 1949, *The Physical Principles of the Quantum Theory*, Dover.

2. For single photons passing through the slits, see Rueckner, W. and J. Piedle 2023, *American Journal of Physics*, 81, 951 ; for electrons, Tonomura, A., *et al.* 1989, *American Journal of Physics*, 57, 117.

3. The Hamiltonian of a system specifies its total energy; the sum of its kinetic energy (motion) and its potential energy (position), in terms of the Lagrangian function which is the difference between kinetic and potential energy. Ref. Lanczos, C., 1970, *The Variational Principles of Mechanics*, 4th Ed., Dover. The *reduced Planck constant* $h/2\pi$ is used because many equations in classical and quantum physics having a *Planck constant* typically also have a 2π factor.

4. Mathematical *forms* relate the maximum precision of a measurement, such as position or energy, to the precision of some conjugate quantity, such as momentum or time. A *form* here means a function defined on a vector space expressed as a function of the coordinates over any basis set.

5. Hermitian (also called *self-adjoint*) matrix means that if the elements of the matrix are complex conjugated and transposed (reflected about the diagonal), the matrix is still the same. For those interested, ponder why such a representation should lead to a real-valued observable.

6. For Fourier series and FFT, see Chapter 21.

7. Schrödinger's cat in the box from en.wikipedia.org., word mixing cartoon generic anonymously from bytesdaily.blogspot.com.

8. Quote from Einstein, A., *Autobiographical Notes*.

9. A unitary matrix is a complex square matrix whose columns (and rows) are orthonormal (orthonormal matrices' columns vectors form a set wherein each column vector has length *1* and is orthogonal to all the other column vectors). They are used to denote the allowed values of quantum mechanical systems that ensure the sum of probabilities of all possible outcomes of any event always equals *1*.

10. For Special Relativity, see Chen, R.H., 2017, *Einstein's Relativity, the Special and General Theories with Their Cosmology*, McGraw-Hill Education (Asia).

11. Regarding quantum mechanics hidden variables and possible incompleteness, see Bohm, D., 1989, *Quantum Theory*, Dover.

12. Gaussian Boson Sampling (GBS) is the probabilistic distribution of bosons from an interferometer. If an optical distribution of N nodes is from M photons injection, then the bosons will produce independent photon measurements. The probability distribution of the sample with the complex matrix is called *permanent* (meaning the product of the rows and columns, and thus the weighting will be $+1$, a very complicated computation_ See

Tacchino, F 2019, *et al.* "An artificial neuron implemented on an actual quantum processor", *Quantum Inf* **5,** 26.

13. IBM and Microsoft make their quantum computers accessible to all through the Cloud, for example, with an IBM quantum account, you can use Qiskit Runtime, via the qiskit-ibm-runtime package, to access simulators and quantum devices. The IBM quantum computer image was taken by Stephen Shankland, see Shankland, S., "Quantum computer makers like their odds for big progress", *CNET*, Dec. 25, 2020. The cylindrical casing holds the processor, and the cables are for input/output.

CHAPTER 28

Industrial Robots

Robot Physicians

PHYSICIANS AND MEDICAL TECHNOLOGISTS have long used medical instruments, such as X-ray machines, blood analyzers, and EKGs, and in 1985, a robotic arm performed a stereotactic brain biopsy with 0.05 mm accuracy.

WatsonQA, in a more useful role than *Jeopardy* or debating, given its supreme ability to know the question from the answer and argue for the best medical advice, could just as well assist physicians or act as robot physicians themselves, provide diagnosis and treatment of patients, while creating lucrative new robot business opportunities for the machine's proprietors.

For example, in the *Jeopardy*-like medical diagnosis game *Doctor's Dilemma*, there was once an answer:

> *The syndrome characterized by joint pain, abdominal pain, palpable purpura, and a nephritic sediment.*

The diagnostic question is (of course) "What is *Hanoch-Schonlein Purpura*?" The same game format used in *Jeopardy* can be used by WatsonQA in the indubitably nontrivial medical diagnostics that requires knowing the medical question when you know the symptoms.

In addition to the onboard memory of the *Jeopardy* machine, an *IBM Watson Health* doctor or physician's assistant machine could be connected

DOI: 10.1201/9781003463542-28

to the Internet and the Cloud, thereby having immediate access to medical information and the latest developments in research. Then quickly assessing the symptoms, diagnosing, and advising treatment, employing background knowledge, and so on that a very competent physician could never match in a lifetime.

IBM Watson Health has programs, for example, in oncology, genomic interpretation, and diabetes management. It employs *Nuance*'s speech recognition software specifically designed for medical terminology to act as a patient interface, so in response to a query, medical information from the Internet and Cloud can be searched in seconds, and questions and answers can use WeChat to organize the information, allowing access to voluminous medical information and expert diagnosis for anyone with a verified connection.

For instance, a physician might say to Watson Health, "My patient has had digestive issues and has lost interest in bowling, her favorite pastime". IBM's *Blue Gene* supercomputer then searches the *Diagnostic and Statistical Manual of Mental Disorders* (DSM) for "lost interest" and classifies that as a symptom of depression, and then scans journals looking for the logical AND of "depression" and "digestive problems" and finds an article on *celiac* disease, an autoimmune disorder. If there are other articles supporting this diagnosis and no clear contradictory evidence found, the physician can order lab tests to confirm or dispel the celiac diagnosis. If confirmed, a gluten-free diet would be advised, and the patient hopefully will soon be happily back at the bowling alley.

To refine the automated diagnoses, Watson Health could assign weights to articles, for example, based on the number of positive citations, and consider for example epidemiological spread factors that increase the probability of a given diagnosis in a given region.

Eighty percent of healthcare data is unstructured, and just like *Jeopardy*'s WatsonQA, Watson Health can read and understand unstructured data by natural language processing to identify, classify, and store clinical information from virtually any source, and just as in every other field today, Big Data, LLMs, GPTs, and WeChat can substantially improve medical predictive analytics with voluminous, and now more structured data.

An artificial neural network trained on massive amounts of medical data is used by Watson Health to classify diagnoses, with reinforcement learning improving its diagnostic accuracy. Again, a human cannot endure 24/7 training and tireless study, and indeed many control experiments have shown that Watson Health's machine learning can be equal to if not superior to diagnoses by human physicians.

Once up and running, Watson Health can perform problem identification and automatically produce a summary of care from a patient's medical record, then after classification of patients with clinical similarity, dynamic patient cohorts can be created for path selection for a given group of patients, with the optimum care paths becoming an integral part of healthcare information available to every practitioner.

For medical research, Watson Health can find information in the medical literature to support new hypotheses and create new diagnostic tools; for example, quickly scanning and reading a complete set of medical literature such as the journal *Medline*, and from there identify documents that are semantically related to the research topic in question.[1]

Notwithstanding, many practicing physicians oppose the use of artificial intelligence for diagnosis or other medical matters, often citing a machine's lack of empathy for the patient (as well as sympathy for the one-upped doctor), and the likely increased liability risk of malpractice stemming from machine misdiagnosis.

Furthermore, having a machine take over medical diagnosis is against a human physician's professional self-interest, and the storied arrogance of some physicians may prevent their acceptance of machine diagnosis, no matter how great the benefits.

Perhaps the debater Harish Natarajan's suggestion of man–machine collaboration would encourage a machine doing the hard work of diagnosis with the human physician deciding on treatment and handling the emotional care of the patient, with augmented machine malpractice insurance, and premiums paid from the higher profits derived from the lower costs of research- and diagnosis-performing robots compared to human physicians.

This is in accord with the first sentence of the Hippocratic oath below, but should the physician's professional and/or pecuniary concerns at the thought of a robot physician taking over patient diagnosis outweigh the duty set forth in the second paragraph below?[2]

> *I will remember that there is art to medicine as well as science, and that warmth, sympathy, and understanding may outweigh the surgeon's knife or the chemist's drug.*
>
> *I will not be ashamed to say "I know not", nor will I fail to call in my colleagues when the skills of another are needed for a patient's recovery.*

When the "colleagues" are physician's assistant machines or very capable *Doctor Robots*, are human physicians violating their Hippocratic oath when they are not receptive to artificial intelligence to help care for a patient?

NYU CANCER DIAGNOSIS

An artificial intelligence automatic recognition and diagnosis of lung cancer from images of diseased and normal tumors was developed by the NYU School of Medicine with tumor image data downloaded from the *Cancer Genome Atlas*, which was prepared by expert pathologists' detailed microscopic examinations of tumors and their diagnoses. This data constituted a training set of 800,000 images from 1,200 cases of diseased and healthy lungs for machine learning. The *Google Inception v3* computer vision convolutional neural network (CNN) learned diagnosis from the image data recognition, and after two weeks of training, the CNN could correctly diagnose tumors at 97% accuracy, better than the three expert pathologists who served as a control group.

Taking a step further, NYU's CNN was asked to extract more than just the cancer diagnosis from the training set images. Expert oncologists cannot discern genetic mutations solely from images of tumors, but rather must read the tumor's DNA sequencing and compare it with the normal DNA of the patient to detect the possible onset of genetic mutations, a tedious and error-prone process.

The NYU automatic cancer tumor diagnostic tool was able to automatically predict the mutational status of a key lung cancer-driving gene with greater than 80% accuracy, and it was found that more training set data could further increase that accuracy rate to far surpass the best human detections of genetic mutations.[3]

ASSEMBLY LINE QUALITY CONTROL

In the complex mass manufacture of liquid crystal displays (LCDs), in spite of almost completely automated production, final inspections of LCD panels were often done visually and if a defect in a panel is found, it can cause an entire production run to be downgraded or just discarded. Moreover, the cause of a defect and when and how it arose is often difficult to ascertain, such that the term for a defect whose cause is unknown has been ruefully called *mura*, a generic Japanese term for "irregular" or "non-uniform" and a word used by car manufacturers as "wasted".

China Star, a subsidiary of TCL, the world's third largest producer of television sets, engaged IBM Watson to develop an *Artificial Intelligence LCD Panel Inspection System* that obviates human visual inspection of defects using computer vision convolutional neural networks to automatically detect, identify, and classify defects from pattern recognition comparison with a database of defective LCD panels and their causes.

The almost completely automated LCD fabrication assembly line has robots at virtually every station, and in addition to a final inspection scan, mounting CMOS sensors on the robots' heads can scan the panels at critical fabrication stations for defects, and if found, the production line can be stopped at that stage and the process adjusted as needed to prevent the defect, and then continued, saving time and avoiding a completely unproductive production run.

The trained AI inspection system can also store new defect data in the database in real-time, and thus improve its defect-detection skills as it works on the assembly line, and so increase the all-important manufacturing yield.[4]

COMPUTER PROGRAMMING

There are many online program-drafting competitions organized by programmers themselves that have attracted developers of code-writing machines to generate human-readable source code given a programming objective in an input–output test.

Automatic programming generation employing *recurrent neural networks* (RNNs), wherein some hidden layers change in response to activation from succeeding layers, can model sequential programming steps to formulate a working program that satisfies the output test objective.

Researchers at Microsoft and Cambridge have developed a *Learning Inductive Program Synthesis* (LIPS) machine called *DeepCoder* that determines the attributes of programming language for a specific task utilizing text recognition to generate a programming dataset of the character-by-character sequences used in the C++ and *Python* high-level computer languages. LIPS learns the probability distributions of attributes for the given programming task employing an artificial neural network, and then guided by the machine-learned input–output mapping derived from the task objectives, searches existing computer programs for program steps consistent with the task objective.[5]

In 2020, *OpenAI* announced its *Generative Pre-trained Transformer version 3* (GPT-3) that could not only write computer programs, but through enormous database training from crawling the Internet and billions of parameters to adjust fit, its algorithms, through reinforcement and unsupervised learning, could compose prose and poetry, and indeed write any text up to 50,000 words.

The dependable, tireless, noncomplaining robot programmers are particularly suited to the Red Bull®-driven all-night sessions of intensive programming, and the unerring placement of all the semicolons and parentheses in C++, and with more training will be able to program faster. Moreover, it will need only electricity for sustenance, with no infrastructure requirements of free Coke®, La Croix®, Red Bull®, nuts, pizza, fried wontons, and ping-pong tables.

The ideas for new program applications at present are the province of human ingenuity, but it looks like that as GPT-3 and its progeny gain programming experience and skill, they will develop entirely new programming techniques and find new uses and entirely new areas for computer programming.

DIGITAL ASSISTANTS

The stand-alone digital assistant can handle personal and business communications, providing information in synthetic speech in response to human speech commands. Its mobile derivative, a service robot also can be a helpmate and companion.

Everyone is familiar with the speech commands understood and responded to by today's personal computers, smartphones, digital assistants, and service robots, which actually do a fairly good job. From the early *Defense Advanced Research Projects Agency* (DARPA)-sponsored continuous speech recognition development and natural language front-end recognizers, to the clumsy adventures of IBM's pioneering but flawed *Newton* and Apple's early malapropistic *Siri*, great strides have been made in *natural language processing* (NLP), and *intention-driven interfaces* such as Nuance, Apple's *Intelligent Siri*, Wolfram *Alpha*, IBM *Watson*, Google *Assistant*, Now, *Nest*, Microsoft *Cortana*, and Amazon *Echo* and *Alexa*, all of which are probing the "known unknowns" of ambiguity and inference in speech, in many languages, and for machine translation.

These artificial intelligence robots are already among us, either helping or bewildering us; will they eventually replace all of the service class for business, government, home, and although they can appropriately respond to human voice commands, will they ever really "understand" human needs and be able to eloquently deliver information like learned humans?

NOTES

1. Ref. IBM Watson Health online.
2. This 1964 revised version of the Hippocratic Oath is used by most medical schools.
3. World Intellectual Property Organization (WIPO) Technology Trends 2019, *Artificial Intelligence*.
4. Regarding LCD fabrication, ref. the author's book Chen, R.H., 2011, *Liquid Crystal Displays, Fundamental Physics and Technology*, Wiley.
5. Balog, M. *et al.* 2017, *DeepCoder: Learning to Write Programs*, ICLR Paper.

CHAPTER 29

Autonomous Vehicles

THE EMPTY SELF-DRIVING CAR cruising the city streets is likely the most dramatic display of the wonders of artificial intelligence. Almost every car manufacturer and new tech company have their autonomous electrical vehicle in development, and emergency braking, lane-changing, and self-parking are all already being widely installed and used every day, but they are just the start of a completely autonomous driving.

The basic concept is based on Norbert Wiener's *adaptive feedback loop*, a system that probes the environment by *Light Detection and Ranging* (LIDAR) diode laser beams, complementary metal oxide semiconductors (CMOS) sensors, radar, radio, and ultra-sonic transmitters and receivers combining to map a spectrum of stimuli in a three-dimensional *point-cloud image*. A feedback response is produced from an artificial neural network's (ANN) supervised and reinforcement learning, on-board computers to collate and compute the image, actuators, and servos to control speed, turns, and braking to maneuver in the diverse environments and changing conditions of typical automobile driving.

The principal component is the LIDAR that microsecond-pulses Class I low-powered and rapidly rotating 905-nanometer-wavelength laser beams for shorter range ($\leq$50 m), and the higher-powered, longer-wavelength 1,550 nm laser beam for longer-range probing ($\leq$200 m).[1]

Any rapidly rotating, wired electronic device cannot avoid the simple problem of the wires winding around itself while spinning, so a rapidly rotating *micro-electromechanical system* (MEMS) mirrors reflect the LIDAR laser scan outwards.

DOI: 10.1201/9781003463542-29

The LIDAR beam is reflected by objects in the environment back to the CMOS sensors and from the time of the returned light signals, the distance of scanned objects is detected, and in accord with the change in wavelength upon reflection by a moving object, the relative motion of scanned objects can be immediately computed using the Doppler effect for light.[2]

However, heavy rain, snow, and fog will distort and blur the LIDAR probe and CMOS sensing, so longer-wavelength radar transmitters and sensors are mounted on the bumpers, and typically 12 ultrasound sensors are distributed around the car body to detect unexpected objects at low speed. But these sensors are subjected to noise and require Fast Fourier Transforms (FFT) to denoise.

It is interesting to note that the pioneering Tesla self-driving cars do not use LIDAR, but depend on eight CMOS sensors and radar for sensing, but the industry-leading BYD depends on LIDAR.

The automobile's location is monitored by global positioning satellites, local street maps, gyroscopes, *visual odometry* (VO), wheel speed sensors, and *inertial measurement units* (IMU) attached to the wheels that detect turns, accelerations, braking, and distance traveled.

All the sensor data, graphics rendering, and computation require large *Graphics Double Data 6* (GDDR6) memory for GPU parallel processing.

On-board computers run *simultaneous localization and mapping* (SLAM) programs to dynamically follow the car in real-time, constrained by the navigation system determining the destination, traffic rules, signs, and so on.[3]

The clustering of all the devices and sensors together with a convolutional neural network (CNN) that has been point-to-point mapped to produce a three-dimensional-like, 60-meter radius *point-cloud map*, as depicted in Figure 29.1.[4]

While navigating the streets of the city and the roads of the countryside, aside from the traffic, the autonomous car must detect pedestrians, obstacles, and particularly anomalous driving and pedestrian behavior, thus it must make immediate responsive decisions based on the probability of the behavior. In this case, strangely enough, it is like the probability prediction of the next word in natural language processing (NLP), and can employ Hidden Markov Models (HMM) and a Markov Decision Process (MDP) algorithms to choose the highest probability (safest) response.

A Bayesian neural network (BNN) probability function can calculate the *joint probability* of given one vector, the result on the value of the other

FIGURE 29.1 Autonomous car point-cloud map.

vector; that is, the choice of a *state-action* maneuver's consequent effect, and after many Markov Chain Monte Carlo (MCMC) simulations are run, optimum responses are determined.

The car's neural network computation has four basic steps: (1) determine the optimum route from source to destination; (2) establish a driving policy from reinforcement learning and the MCMC simulations; (3) when faced with an anomalous situation, choose the response based on the optimal joint probability MCMC simulation results; and (4) follow the optimum driving policy with exceptions for exceptional circumstances.

From the simulated 3D point-cloud map DCNN feature extractions, the autonomous car can operate based on its *receptive field* in *feature space*; that is, other cars, traffic signs and signals, pedestrians, road features, and so on. Daily drives to the office typically will have the same environment, and in a recurrent neural network (RNN) certain features will always reappear and can be stored as landmarks and surroundings with "hard weights".[5]

From training sets of actual driver experience responding to many different situations, the autonomous car learns how to drive, and clearly, the more training data, the more optimal the responses, and the more real-time or simulation driving experience, the more proficient the driving. Thus, the self-driving car learns to drive like a human from the bottom up, and gains greater skill by driving more and encountering more different situations.

Having undergone millions of Q-learning *state-action pairs'* reinforcement learning simulations, just as the AI Video Gamer, self-driving cars will continually update their driving policy using the Bellman equation for immediate action decisions, and gradually improve their driving skills.

Self-supervised learning can be used for identifying unusual or unexpected features or events, such as a jaywalking pedestrian, double-parked cars, or abrupt lane changes. The sensor data can be *clustered* to identify different types of road surfaces, specific traffic conditions, temporary detour signs and arrows, and the like.

Social Darwinism is also used to evolve better self-driving cars; that is, *population-based training* (PBT) takes the skills of the best drivers (both autonomous and human) and programs them into the car's computer in a *fittest driver selection*; in this way, all self-driving cars will be the best drivers in society.

The autonomous car will communicate through 5G, 6G, Internet of Things (IoT), and communication satellite to an on-board computer or central autonomous car driving center, and thereby integrate the whole driving experience for the passengers' comfort, safety, and route scenery, if desired.

Self-driving cars can also drop you off at your destination, go find a parking space, and upon a mobile phone command, return or find you elsewhere.

New driving experiences are also being introduced, such as BYD's 180° turn in place, driving in water, and the Xpeng AeroHT, which can fly in the skies (but "flying cars" will be used mostly just to gently rise above rough roads and terrain, and skim smoothly on its way).

Particularly on highways, the swarm of cars can have a group characterization and intelligence to avoid accidents. A *swarm* of self-driving cars will all share in and contribute to the entire group's driving skill set, much like a swarm of bees can unerringly home in on their target, and in the process never bump into each other, or impede the progress of the whole.

Of course, individual cars on their own particular routes will deviate from the swarm formation, but automatic lane changes and detectors at the exits can take over the car guidance. Turning into a parking space will be controlled by the navigation and self-parking system and park itself in the garage. Perhaps best of all, the reckless and drunken driver will be immediately spotted by the central autonomous car system, his license revoked until he purchases, and is only allowed to be a passenger in a self-driving car.

As for internal combustion driver diehards, there can be a separate "human driver's lane" on the highway, but the boredom and dangers of driving on highways will lead even the most avid motorists to eventually join the swarm and drink coffee while playing with their mobile phones.

During the development of self-driving cars, just as in the experiment and testing of any new technology there will inevitability be mishaps, and critics will condemn the autonomous car as a technological danger.

However, statistics show that, for instance in the United States, there are nine deaths and more than a thousand serious injuries *every day* caused by *driver distraction*. A self-driving car is programmed only to drive and cannot be distracted, either by emotion or mobile phones (where calls can be automatically answered ready for response). Furthermore, studies show that more than 90% of accidents are due to driver error; driverless cars eventually will reduce those errors to near zero.[6]

In the beginning, the greatest obstacle to self-driving cars was in the handing over control to a machine in a dangerous environment. However, research surprisingly showed that even the most obstinate "control freak", after overcoming initial wariness, sits in the autonomous car for about 15 minutes, and seeing that nothing untoward has happened, calmly settles in; after all, it is the same as taking Uber, a taxi, or bus.

Perhaps the only outlier will be the Italian super sports car afficionados who love the sound of an internal combustion engine at acceleration to full throttle, but he may be nonplussed by the fact that a mid-level electric car can beat the Ferrari in 0 – 100 km/h acceleration, and high-performance car sounds can be programmed if desired.

The road map for autonomous vehicle development classifies the standards and the degree of autonomy. The Society of Automotive Engineers (SAE) has established five levels of progression for the automated driving automobile industry:

(1) Shared Control Driver Assistance: for example, adaptive cruise control, automated parking, and lane-keeping assistance;

(2) Automated Driving but Monitored Intervention: for example, unforeseen encounters such as roadblocks where the driver must take over control;

(3) Self-Driving and Vehicle Warning for Intervention: for example, the autonomous car will not automatically respond but warn the passenger;

(4) Complete Self-Driving in Specific Environments: for example, on designated roads where the passenger need not intervene; and

(5) Complete Self-Driving in any Environment: for example, where the passenger relinquishes all control to the autonomous car, which may not have a steering wheel, but likely has an emergency brake.

The level in advanced countries has at best reached level 4, currently mostly for shuttle buses, and eventually highway trucking, but it will take some time before level 5 is realized for all cars.

But AI features integrated with the mobile phone or smartwatch or other IoT and many other appsare available in your car, for instance, voice commands, video calls on car screens, smart cameras, remote door locks, household appliance commands, conversation with a digital assistant, and so on, and these are already available and customizable.

NOTES

1. Lasers are classified 1 – 4 according to wavelength and maximum output power damage to the eye. The visible wavelength range of light is 400–700 nm, so both the 905 and 1,550 nm are not absorbed by the retina, but high-powered laser light may cause corneal damage. Electromagnetic waves interact with matter by resonating with lattice molecules in the matter with comparable lattice dimensions. Short wavelength blue light will interact and be absorbed or reflected, long wavelength radio waves have wavelengths longer than most matter lattice dimensions and therefore are able to pass through obstacles unimpeded. The lasers are in conformance with the US Food & Drug Administration's eye safety standard and the International Electrochemical Commission's 60825 performance standard. The complementary metal oxide semiconductors (CMOS) sensors are transistors that sense light.
2. The Doppler effect is the change in wavelength (or frequency) of a wave source moving away (increasingly lower pitch for sound and redshift to longer wavelengths in the electromagnetic wave spectrum) or towards an observer (higher-pitch sound and blueshift to shorter wavelengths).
3. SLAM is probably not the best acronym for self-driving cars.
4. Point-cloud image by Graham Murdoch, *Popular Science*, personal communication from the University of Michigan, developed by the CSE Department.
5. For "hard weights", see Chapter 23.
6. Center for Disease Control and Prevention, 2019 cdc.gov/motorvehicle-safety/distracted driving.

CHAPTER 30

Exoplanet/Exomoon Astronomer

F ROM DECEMBER 18 TO 28, 1995, the Hubble Space Telescope had turned to make a ten-day-long exposure covering just 3×10^{-8} area of the Northern Sky. Astronomers, expecting an empty black void, were astonished to find 3,000 galaxies in the digital *Hubble Deep Field* image (Figure 30.1), which if extrapolated to cover the whole sky of the Earth means that the visible Universe then was estimated to have at least $3,000 \times 1/(3 \times 10^{-8}) = 1 \times 10^{11}$ galaxies, a number that since has been raised to 2×10^{12} in 2023 by further observations and calculations.[1]

There are now estimated to be an astounding $3 - 7 \times 10^{22}$ stars in the visible Universe. Our galaxy, the Milky Way, has about 4×10^{11} stars, of which 22% are similar in size to our Sun, and of the estimated 6×10^9 planets, observations of over 5,500 planets have been made by 2023.

Astronomers believe that like our Solar System, during stellar evolution, there should be iron-rich rocky planets such as the Earth orbiting in the habitable zone, suitable for human development, and estimate there should be about 6×10^9 Earth-like planets in the Milky Way Galaxy alone.[2]

The search for extraterrestrial life is one of humankind's most compelling pursuits. Presently over five thousand planets orbiting relatively near stars have been discovered, more than half of which were made by the Earth-trailing heliocentric orbiting *Kepler Space Telescope* (KST) launched in 2009.

DOI: 10.1201/9781003463542-30

FIGURE 30.1   Hubble Deep Field image.

Kepler's digital cameras observed 530,506 Milky Way stars, and the brightness-over-time light curves (*light curves*) calculations revealed the tiny periodic changes in paths of stellar brightness caused by exoplanets transiting its star.

To find planets at light years' distance away requires following the dim trace of a putative planet across its star with a sensitive and stable telescope to focus on the *exoplanet*. Unfortunately, in 2012 and then in 2013, two of the Kepler telescope's four stabilizing directional reaction wheels failed, and the telescope could not hold a stable pointing position, resulting in less precise and very noisy observational data.

KST's remaining two reaction wheels, with the help of a small stabilizing rocket, could roughly control the telescope, and during the so-called *KST Second Light*, KST could be oriented to receive uniform solar radiation pressure such that it could turn to follow a star (Figure 30.2).[3]

However, as long as KST was still looking, there should be exoplanets passing through its field of view at the same rate as before, so Anne Dattilo at the University of Texas designed *AstroNet-K2*, a deep artificial neural network that after training on known exoplanets, could systematically remove the instability and noise from KST's signals, and not only reveal new exoplanets, but also find exoplanets in the old observational data *that even experienced exoplanet astronomers had missed*.

The most difficult *denoising* of the exoplanet's transit came from the small disturbance in the light curves resulting in false positives, but after further supervised training on the exoplanet-hunting data, AstroNet-K2

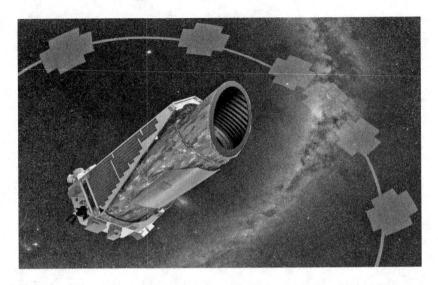

FIGURE 30.2   Artist's rendition of Kepler Space Telescope second light.

could perform with 98% accuracy, equal to, or surpassing expert experienced astronomers.

Furthermore, from December 2016 to March 2017, as Mars passed through the crippled KST's field of view, its direct and scattered light obfuscated any exoplanet signatures, but AstroNet-K2 heroically discovered two exoplanets by removing all the reflective glare of the Red Planet.[4]

One detected exoplanet was a super Earth-sized, volatile-enveloped "puffy" planet whipping around a Sun-like star with a 13-day period and a surface temperature of 750°C, a little too hot for humans, but fast-paced, heat-loving beings would love the quickly-passing seasons of super-tropical weather.

The second exoplanet was also super Earth-sized, but with an even shorter 3-day period that would truly make "the hours pass like minutes", and a surface temperature of 1400°C, hot enough to melt aluminum let alone humans. One wonders what any beings on these exoplanets would look like ....

The first Earth-sized exoplanet was discovered in 2015 also by the impaired Kepler Space Telescope. Prosaically named *Kepler-452b* is orbiting in the habitable *Goldilocks* zone around *Kepler 45*, a star 10% larger than our Sun, with an exoplanet orbiting in the habitable zone about 60% bigger than Earth, but with an orbital period of 385 days, almost the same as an Earth year, and an equilibrium temperature of −0.8°C (compared with Earth's −0.18°C).[5]

Because Kepler-452b's star is older than our Sun by about 1.5 billion years and considerably brighter, Kepler-452b is warmer than our Earth, but water appears to be present as in oceans, and assuming a suitable atmosphere and pressure, it is amenable to an H_2O triple-point. If there indeed is liquid water then *thermophilic bacterium* may exist and gradually evolve to become *eucaryotic cells*, which are the precursors of Earth-like microorganisms that are the building blocks of life.

However, at about two times the size of Earth, its surface gravity is also about two times greater, so any *Earth 2.0* animals would not need much fur, and its humanoids would be tanned, stocky, and very muscular.

If these super-strong "Kepler 452b-ings" are intelligent enough to develop themselves or their robots to travel the 1,402 light years distance to Earth, in the absence of spacetime wormhole travel, it would still take too long to have any meaningful communication.

At the speed of the fastest rocket as of September 2023, the *Parker Solar Probe*'s 59,000 km/h, it would take approximately 30 million years to get to K-452b, and even at close to the speed of light, it would take at least 2,804 years for K-452b beings to communicate, reach our Earth, and colonize us, so we are safe for now, unless wormholes really exist somewhere in the Universe.

Earth's closest star is the red dwarf *Proxima Centauri*, only 4.2 light years away, and has an Earth-like exoplanet *Proxima Centauri b* with mass just 1.27 times that of Earth mass, and in a habitable zone, but with an orbital period of just 11.2 days, something that would truly "make the years pass like days", and if there is no ozone layer to block the ultraviolet glow of the red dwarf *flare star*, the hydrogen, oxygen, and nitrogen necessary for an Earth-like atmosphere would have been stripped away.[6]

If the water is underground, the thermophilic bacterium could evolve into beings and with the remaining nitrogen oxides (NO_x) possibly reacting with plants' volatile organic compounds (VOC) and $C_2H_2C_{12}$) to produce an O_3 (ozone) layer and beings could come out of the water possibly with gills, and be very good swimmers in a movie setting like *Water World*.

For all that biochemistry to happen, it seems to require too many factors and coincidental happenstances, but the 13.7×10^9 years of the Universe with at least 10^{22} stars and an estimated 6×10^9 Earth-like exoplanets, there is time and room for almost anything to happen, as it did on Earth.

If the aliens' technology can produce a close-to-the-speed-of-light rocket, they can be transported in suspended animation, and with relativity's time dilation, they will age more slowly, and may be able to reach Earth in about ten years.

This dire possibility of a close encounter with exoplanet beings prompted astronomers at MIT and the Carnegie Institute for Science in 2017 to release two decades of data, and the software and an online tutorial to analyze that data, and called on amateur astronomers to help with observations of the more than 1,600 stars within 325 light years from Earth in the hope that crowd-sourcing "fresh eyes" would find new nearby exoplanets, not only for the scientific adventure but also to gain time to prepare for eventual alien landings on Earth.[7]

It is clearly incumbent upon humans to keep looking for exoplanet life and communicate if possible. The Kepler Space Telescope could have helped, but its remaining two reaction wheels finally ran out of fuel and it was officially retired on October 30, 2018. Fortunately, a new space telescope, the *Transiting Exoplanet Survey Satellite* (TESS) had been launched on Elon Musk's Space X Falcon 9 rocket on April 18, 2018, and is continuing the epic search for extraterrestrial life.

Advanced extraterrestrial beings would no doubt first communicate and send messengers before landing *if* they meant no harm, otherwise we should prepare our defenses to alien attack.

Well before such an invasion takes place, for our very survival, Earthlings must develop artificial intelligence to supplement our meager native intelligence and send our robots to get to them first. In this sense, we will need AlphaZero's inference engine intelligence not just to amaze us, but more importantly to save us.

AstroNet-K2 and later-developed astronomical machine learning could automate much of the work of exoplanet hunters, working tirelessly at any time and place under any conditions, and without the biases that humans might have, particularly in the urgency to find Earth-like planets peopled with human-like beings in this most glamorous, and possibly most *critical*, field of astronomy.

It would take 8.42 years between transmitting to and receiving electromagnetic wave messages from *Proxima Centauri b*, and the *Search for Extraterrestrial Intelligence* (SETI) multichannel analyzer is continually sending and looking for such messages, but so far has only recorded many false positives.[8]

The more advanced TESS will continue KST's exoplanet hunting aided by artificial intelligence supervised learning like AstroNet-K2, reinforcement and self-supervised learning, and to start with, analyze 10,803

Threshold Crossing Events (TCE) light curves that indicate the possibilities of exoplanets.[9]

Together with the Australian telescope array in the Southern Hemisphere *Minerva*, the James Webb Space Telescope's infra-red *Wavefront Sensing and Control* charge-coupled device (CCD) launched in 2021, and China's *Five Hundred Meter Aperture Spherical Radio Telescope* (FAST) will continue the exoplanet hunt, with the help of artificial intelligence.

EXOMOON DISCOVERY

Astronomers have long believed that since for instance Saturn alone has 82 moons, large and small, it would be entirely reasonable that other planets around other stars would also have moons. Researchers sifting through the light curve data from 284 exoplanets found by the Kepler Space Telescope indeed spotted the telltale secondary brightness dips following an exoplanet's signature dip while transiting the star.

The light curve showed that exoplanet K-1625b has a moon as massive as the Earth and four times its diameter, K-1625b I and was confirmed by the Hubble Space Telescope, but there was controversy over whether it was an exoplanet or exomoon.

This issue could use artificial intelligence to settle. That is, perhaps an exomoon is a latent factor in collaborative filtering of the light from K-1625b I ("inferences from the inference center"), and an AI classifier such as a support vector machine (SVM) or a generative adversarial network (GAN) could discriminate a planet from a moon. Furthermore, the massive data regarding exoplanet and exomoon discoveries could be assembled in a specially designed large language model (LLM) and a generative pre-trained (GPT) with billions of parameters to analyze the possibilities. At the very least, these artificial intelligence models could improve the resolution of all the telescope images.

In July 2019, researchers using the *Altacama Large Millimeter/Submillimeter Array* (ALMA) in Chile inferred from fuzzy splotches in millimeter wave patterns 370 light years away that the young planet PDS 70 c orbiting the T Tauri star PDS 70 has a circumplanetary disk one-fourth the mass of the Earth's moon. Further studies have found numerous exomoons not so far from Earth.

Computer vision pattern recognition by a deep convolutional neural network (DCNN) and data analytics could be employed on the images of giant radio telescope arrays in Chile and China to search for more exoplanets and exomoons.[10]

AstroNet-K2 could only spot the type of exoplanets that it had learned to recognize, but with reinforcement learning and self-supervision it could begin to think for itself to discover exoplanets with different signature characteristics, and like AlphaZero and the AI Video Gamer in their pursuits, could outperform even the most expert human astronomers in finding exoplanets and exomoons.

In doing so, the AI astronomer will be assisting humankind in what is perhaps its ultimate undertaking, finding other intelligent beings in the Universe. It has been estimated that in our Milky Way Galaxy alone, almost every star has some orbiting planets and there are therefore more than one trillion exoplanets and even more exomoons to be discovered.

Thus, humankind can either create intelligent beings here on Earth in the form of thinking robots, or find them in our own galaxy as developed by superior beings. For the 1 trillion exoplanets developing during the 13.7 billion years of the Universe, taking the Earth as an instance, intelligent beings will surely evolve on some of those exoplanets, and like us, they will surely develop artificial intelligence, and if they are biologically more robust and can devise a warp speed spaceship, they may venture out to a close neighbor, such as our Earth. In this instance, all our geopolitical conflicts and wars will become mere trivialities.

In the 2 trillion galaxies in the Universe and 1 trillion exoplanets per galaxy, 2 trillion trillion (2×10^{24}) exoplanets have had and will have billions of years to develop some form of life, so by the sheer dint of numbers and time, there is little doubt that intelligent beings and their intelligent creations are on exoplanets and exomoons in the far reaches of all of the galaxies.

In this sense, the development of intelligent robots appears inevitable in some Galaxy, and in spite of its threat to humankind's dominance on Earth, it is incumbent on humans to do the same, if for no other reason than our own or our electromechanical progeny's advancement to maintain relevance in the Universe.

NOTES

1. The number of stars is calculated by determining a galaxy's mass as it rotates and changes its light spectrum, and at best can be only a very educated guess from studies of the shapes and inclinations of the galaxy. Rf. *Space.com*, February 12, 2022, for details of the calculations.
2. The number of galaxies is determined by telescope deep exposures of a sector of the sky and then extrapolation over the entire sky. The 1995 Hubble

Space Telescope Deep Field ten-day exposure by Robert Williams and subsequent deep-field exposures by the Wide Field Camera 3 on the HST and the Subaru and Keck Telescopes taking exposures of different wavelengths together with the Doppler effect redshift provide depth, and the total number of galaxies could be calculated from the galaxies' known Gaussian and Power mass distributions. Conselice, C.J., 2017, June, *Our Trillion-Galaxy Universe, Astronomy*. Since then, there have been many more observations and extrapolations.

3. Hubble Deep Field image and Artist's rendition of the Kepler Space Telescope image from NASA whereby all images are in the public domain.

4. A. Datillo, *et al.* 2019, *Identifying Exoplanets with Deep Learning II: Two New Super-Earths Uncovered by a Neural Network in K2 Data*, arxiv.org/abs/1903.10507.

5. A. Teachy, *Evidence for a Large Moon Orbiting K-1625b*, Science Advances, 03 October 2018. This star was only 1 million years old; the light brightness of T Tauri stars change over time. An additional 20 moons of Saturn were observed at the Mauna Kea Observatory in 2019.

6. A *flare star's* magnetic energy causes volatile brightness changes over the entire electromagnetic spectrum from x-rays to radio waves.

7. Call for crowd-sourced astronomers, see *Astronomy*, June 2017, p. 13.

8. One of the authors was at Stanford University when young Harvard professor Paul Horowitz was setting up and operating a multichannel analyzer to send and receive signals from outer space. One often wonders what one would think or do if we actually received a message from aliens.

9. Rf.s Ofman, L., *et al.* 2022 February, *New Astronomy*, vol. 91, NASA Exoplanet Archive, January 21, 2023.

10. Rf. A. Isella, *et al.* 2019, *Detection of Continuum Submillimeter Emission Associated Candidate Protoplanets*, Astrophysical Journal Letters, Vol. 879, No. 2. Unfortunately, the Arecibo radio telescope in Puerto Rico collapsed in December 2020 just after de-commissioning. Two times the size of Arecibo, China's Five hundred-meter (diameter) Aperture Spherical Radio Telescope (FAST) began operations in 2019, and will continue humankind's searches in deep space.

CHAPTER 31

Protein Folding

IN LATE 2019, THE *coronavirus*, Covid-19, spread rapidly across the world, and governments and public health agencies were overwhelmed by the virulence of this new contagious threat to humankind, and were determined at all costs to control its spread.[1]

The first objective was to determine its propagation, so that regional medical resources could be prepared, but since this was a novel virus, there was insufficient data to model the pandemic's spread. Thus, for example, a *Kalman filter*, as now used in predicting the position of a self-driving automobile on a journey using trip data, measures the current motion state vectors and estimates their uncertainties, then updates and weights the data as more information from multiple sources is collected.

The Kalman filter would require data to predict the virus' spread pattern and trend, Taking advantage of modern social media, among others, a Canadian company called *BlueDot* continuously collected online disease-related news, official reports, social media mentions, and even air-traffic data, and then cross-referenced the data with the National Institutes of Health and Global Microbial databases.

Strange as it may seem, deep recurrent neural networks (DRNN) such as Hidden Markov Models (HMM) designed for natural language processing have been employed in epidemic spread models by the prediction of sequences (such as words in a sentence) based on previous data and its occurrence characteristics.

A natural language processing (NLP) algorithm correlated the data through the interpretation of a *focal word* (for instance "fever") that influences the interpretation of other words, thereby helping to identify

DOI: 10.1201/9781003463542-31

Covid-19's distribution through *text-mining*. This allowed BlueDot to predict 127,000 cases for March 30, 2020.

Google Brain's simple contrastive learning self-supervised algorithm (*SimCLR*) together with a downstream classifier was applied to identify the Covid-19 virus in medical images of lungs.

However, there are more than two hundred viruses known to infect human beings, each with different infection mechanisms, behavior, and response to treatments and vaccines. The second public health task was to understand the coronavirus infection mechanism.

When the *severe acute respiratory syndrome coronavirus-2* virus enters the body, mostly through the mouth or nose, it infiltrates healthy cells by binding to the cell receptors on the surface of human cell by means of the *protein spikes* studded on itssurface. Infiltrated by the virus, the cell then replicates its RNA, as well as the structural proteins needed to assemble new viral particles, which are then released throughout the body.

Since there was no known cure for the new *Covid-19* virus, it was necessary to find a vaccine to control the pandemic. Universities and research institutions all over the world pursued at least eight different types, including *inactivated viruses* and DNA and RNA vaccines.

INACTIVATED VIRUSES

A virus interacts with the human body host cells through entry into the *angiotensin-converting-enzyme 2* (ACE2) receptors and combined through endogenous receptors in the *receptor-binding domain* (RBD), infiltrates the host cells, and from there transmits the virus through the body.

The immune system activates *cytotoxic T-cells* that destroy the infected cells, and *helper T-cells* that send signals to the immune system to fight the infection with antigens produced by the cells themselves. The antigens bind to *B-cell receptors* to signal the activation of the immune system to produce antibodies to combat the virus' spread in the body.

An inactivated virus is grown from the virus itself, but inactivating or killing it using physical means or chemical reactions so that injecting the inactivated virus into the body will not cause infection, but will stimulate the immune system to produce antibodies to fight the infection.

However, the constructing, producing, and clinical testing for inactivated viruses generally take years to complete and present real dangers of engineering infection. But through intense effort, the first inactivated viruses were effectively administered in about a year and a half and began to significantly reduce coronavirus infections.

PROTEIN FOLDING

A more basic approach was to understand the molecular structure of the coronavirus protein conformation, which dictates the protein function. The deconstruction of a virus to reveal molecular structure and create a vaccine protein is called *protein folding*, and is an exercise in finding a protein conformation that does what you want it to do, here binding onto the Covid-19 proteins to inhibit its ability to enter the ACE2 receptors, effectively preventing a virus from having any physiological effect except to activate the body's immune response.

The idea was to produce a folded protein to directly attack the coronavirus from its basic amino acid sequence structure. However, a typical protein can have 10^{300} different folding conformations and combinations thereof. Furthermore, constructing a three-dimensional protein structure was hitherto primarily performed by freezing a protein into a crystal-like structure and utilizing x-ray crystallography to examine the folding process in an instrument-rich and time-consuming procedure that ultimately could only produce a single protein fold conformation candidate.

Artificial intelligence thus was enlisted for one of biology's grand challenges, moldeling the three-dimensional structure of proteins from their amino acid sequences. Instead of using inactivated viruses, DNA and RNA genomes can mimic a part of the virus' genetic sequence to either block the protein's ability to enter the ACE2 or prompt host cells to produce antigens that trigger an immune response.

The first step was to deconstruct the coronavirus by *Cryogenic-Transmission Electron Microscopy* (C-EM) to find parts of the virus to which a folded protein can bind to inhibit the coronavirus infection and activate antibodies from the immune system.

A Covid-19 virus is depicted in Figure 31.1; with the characteristic *spike proteins* looking like plugs or handles arrayed on the surface, hence the adjectival appellation *corona*.

DeepMind's *AlphaFold* neural network was employed to predict the three-dimensional shape of the coronavirus based on its genetic sequence; then machine learning was enlisted to predict which parts of the virus can be candidate targets for doing what you want it to do, based on a training set data of known pathogens. There are, however, tens of thousands of possible targets on the virus.

It was found, reasonably enough, that the spike proteins arrayed on the surface of the virus were the best targets; the problem then was to design the folded protein to block the spike proteins and trigger an immune

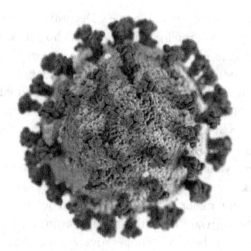

FIGURE 31.1   Covid-19 coronavirus.

FIGURE 31.2   Covid-19 coronavirus protein folding.

response, and not by the way, produce an *unfolded protein response* (UPR) that could foment another disease.

For protein folding, a Hidden Markov Model (HMM) was trained with amino acid sequences and secondary protein structures whose three-dimensional conformation is known, and use them for protein fold classifications.[2]

There are four stages of folding: *primary* (amino acid sequence held together by peptide bonds), *secondary* (protein folded as an alpha helix held together by hydrogen bonds in the direction of the helical axis, or beta-pleated sheets held together by hydrogen bonds in an S-shape), *tertiary* (protein folded into a 3D conformation held together by noncovalent interactions between side groups), and *quaternary* (a single peptide bond to other peptides) as shown in Figure 31.2.[3]

Artificial neural networks can perform protein folding typically in a matter of hours, and new techniques are being developed to reduce the time to seconds. For example, Google's DeepMind *AlphaFold* won the *Critical Assessment of Protein Structure Prediction* (CASP) competition in 2018 by a sizable margin over other competitors, Harvard Medical School protein folding algorithm source code was publicly available on GitHub; Baidu produced a *Linearfold* mRNA protein-folding algorithm, the Washington University (St. Louis) *Folding@Home* open source computer network ran at *exaflop* (10^{18} ops) speeds for crowd-sourced protein folding, and IBM, Google, Amazon, and Microsoft have all made their computer power available to institutions for research on vaccines against the coronavirus.

Universities and major tech companies all joined in the fight for constructing the software algorithms, and hardware, such as the Intel-Argonne National Laboratory cooperative *Aurora* supercomputer took part in calculating the protein folding conformations.

As a first step, AlphaFold employs a deep neural network (DNN) to extract features from a training dataset and then searches for plausible protein structures having those features. It compares a protein's amino acid sequence with similar ones in the training set to find pairs of amino acids that appear in tandem, but do not lie next to each other in a chain, implying that they are positioned near each other in a folded protein conformation in a process it called *Multiple Sequence Alignments*.

The DNN was trained to take the pairings and predict the distance between them in the folded protein. Then the predictions were compared to precisely measured distances in known proteins, thereby enabling realistic guesses on how the proteins may fold.

A parallel-running DNN would predict the angles of the joints between consecutive amino acids in the folded protein chain. The two parameters then could be combined to produce a folded protein structure designed to perform a desired protein interaction such as binding and blocking the spike proteins of the coronavirus.

This theoretical protein-folding design process, however, can produce structures that may not be physically possible. Thus, the DNNs are trained on actual protein structures and the machine learning Cost Function was minimized by gradient descent and backpropagation to come closest to a folding arrangement consistent with the predictions of the amino acid sequences that were produced in the first step, thereby producing a physically viable antiviral protein.

The DNA-based process of developing an antiviral vaccine mimicked a part of the coronavirus' genetic sequence that gives a preview of the virus in order to generate antibodies, but not cause the disease itself, instead readying the immune system to attack any actual infection from the virus.[4]

Harvard Medical School's one-step protein folding algorithm employed a deep recurrent *geometric* neural network based on natural language processing techniques, and trained on a dataset of amino acid sequences mapping to known (and therefore possible) protein structures where the end-to-end sequence-to-structure procedure was performed in milliseconds. The code was publicly available on GitHub in hopes of wide-range dissemination and crowd-sourcing access for further insights and ideas.

A conveyor of the information in DNA to instruct the cell to make proteins from the amino acid sequence is called *messenger RNA*; mRNA is synthesized by complex RNA molecules using the nucleotide sequence of DNA as a template in a *ribosome* factory in the cell nucleus.

Simply stated, after the protein folding has produced the folded protein vaccine, it is injected into the body and a small part of the virus template, the mRNA, thus becomes a "messenger" informing the cells in the body on how to generate the virus protein, including the coronavirus' spike protein.

When the body has been infiltrated by the virus, the cells that have been informed can recognize the coronavirus, activate its immune system, and then naturally produce antibodies. The mRNA is then broken up by the cells and dies, but the body's immune system also destroys the spike protein, and the affected cells will "remember" the coronavirus' presence and produce antibodies. Differently from inactivated virus vaccines, the mRNA, its mission accomplished, dies naturally.

The heroic deeds of the mRNA cannot help but evoke the Greek saga of Pheidippides (Φιλιππίδης), who ran the 40 km to Athens to report the victory of the Greeks over the Persians at the battle of Marathon, and after reporting the good news, he dropped, and died of exhaustion, but his feat heralded a major track-and-field endurance competition.

The mRNA breakthrough was by Penn Medicine which enabled the coronavirus vaccines commercialized by, among many others, Moderna and Pfizer-BioNTech (who likely will win the prize for the clunkiest name for a product, and so was simplified to "BNT" in many countries).[5]

A viral pandemic will eventually peter out naturally because two things happen: (1) infected people produce antibodies that destroy the virus infection, and recover, and (2) infected people do not recover and die, depriving the virus of host cells to live in and from which to spread.

The presence of antibodies can be used as a test for the disease, the plasma containing antibodies (but not red blood cells) can be infused in patients for convalescent serum immunotherapy, and the antibodies can prevent infection. Successful vaccination programs and herd immunity have caused the coronavirus to wither away.

However, this is the rub, the virus can *mutate*, which may break out to foment another serious epidemic. The defenses to viral mutations are new antibody plasma and mutation-specific protein folding, and with more and more data available, data-dependent artificial intelligence machine learning can be marshaled to design new treatments and vaccines much faster and more effectively.

Certain parts of a virus' surface proteins have a high turnover rate producing mutations. Over the past year, tens of thousands of coronavirus samples from patients around the world have been genetically sequenced and uploaded into the *Global Initiative on Sharing All Influenza Data* (GISAID) hosted in Germany. AI algorithms compare those sequences to find which segments of the virus change frequently and which do not, to help identify mutation hotspots. Then (hopefully) the same protein-folding processes can be performed to counter the mutations.[6]

However, it is likely that in the future, just like influenza shots, the population may need yearly coronavirus mutation vaccine jabs.

NOTES

1. Because of its "corona" of protein spikes, it is called *coronavirus* (Covid-19). One hundred years earlier beginning in 1918, the fatal worldwide H1N1-A virus, misnamed the "Spanish flu", in just two years infected nearly one-third of the human population (about five hundred million people) and killed 50 million. There are still 200 known viruses for which there is no vaccine, and every virus has its own structure and infection mechanism, and can further mutate. Medical science has successfully controlled diphtheria, measles, mumps, German rubella, polio, yellow fever, cholera Japanese encephalitis, meningitis A, typhoid, dengue, rabies, and smallpox, but no effective vaccines have been found for malaria, HIV, Zika, West Nile, Lyme, hepatitis C, and many others known and as yet unknown.

2. Coronavirus and protein-folding images are Wikipedia commons open source images adaptations.

3. An *amino acid* (*α-amino carboxylic acid*) is an organic molecule made up of a basic amino group ($-NH_2$) and an acidic carboxyl group ($-COOH$); an organic *R group* – a side chain – attached to the α-carbon atoms of the amide spine displays the charge and polarity of the amino acid and thus determines the chemical characteristics that promote biological interaction.

A *peptide* is a short chain of amino acids and an *enzyme* is a protein and a biocatalyst that converts molecules (substrates) into different molecules (products) in the catalytic metabolic processes necessary to sustain life.

4. *Deoxyribonucleic acid* (DNA) is a double stranded helical molecule that resides in the nucleus of a cell and carries genetic instructions for the development, functioning, growth, and reproduction of all known organisms and some viruses. *Ribonucleic acid* (RNA) is a single stranded molecule synthesized in the nucleus but residing in the jelly-like cytoplasm inside the membrane of a cell; it codes, decodes, regulates, and expresses genes. Both are just chemical compounds, neither is by itself alive.

5. The 2023 Nobel Prize in physiology or medicine was awarded to biochemist Katlin Kariko and immunologist Drew Weissman for discovering a way to deliver mRNA into cells without triggering an unwanted immune response by swapping one molecule (uridine) for another molecule in the genetic material (pseudouridine) that led to the development of the mRNA vaccine. As of 2023, the vaccine has been administered more than 13 billion times.

6. Ref. E. Waltz, IEEE Spectrum September 29, 2020.

CHAPTER 32

Intelligence

A RTIFICIAL INTELLIGENCE MACHINES HAVE bested humans in board games, video games, debating, manufacturing, astronomy, and even poker; and wider, more diverse LLMs and GPTs can interact, write, translate, compose music, create images, and generally perform almost all human intellectual endeavors. But are these artificial neural networks really doing the thinking that requires intelligence, or rather just responding to commands from humans and sifting through information compiled by humans beings?

"Intelligence" has been variously defined, perhaps the most general being,[1]

The ability to acquire and apply knowledge and skills.

All the game-playing and working machines described earlier easily satisfy the elements of this definition. So cognitive psychologists have added,[2]

The ability to perceive or infer information, and to retain it as knowledge
  to be applied towards adaptive behaviors within an environment
or context.

An artificial neural network does exactly just that, so AI machines are intelligent according to this definition.

In an apparent attempt to delve deeper into *thinking*, the definitions of intelligence above have been buttressed with,[3]

DOI: 10.1201/9781003463542-32

understanding, reasoning, critical thinking, planning, emotional
knowledge, creativity, self-awareness, and consciousness.

First, the easiest: a definition of "to understand" is "to form an opinion or reach a conclusion through reasoning and information"; a definition of "reasoning" is "the thought process that leads to solutions of problems". Artificial neural networks perform both in machine learning from a training dataset (information), and by the parameterization of new data, can recognize new data that presents a problem to be solved.[4]

As for "planning", Deep Blue's tree search and progressive deepening, and the discount factor γ in AlphaGo's reinforcement learning are "look ahead" plays that evince planning. What is meant by "emotional knowledge" is anybody's guess, but if it means knowledge of emotions as opposed to logic, then a CNN's controversial interpretation of mood from facial expression and body language could satisfy this rather abstruse criterion.

As for "critical thinking" and "creativity", AlphaGo's Game 2 "shoulder hit" black 37 had a very low algorithmic probability, surprising even its own programmers, but was played by AlphaGo nevertheless, thus meaning it had been *critically* thought out, and it was lauded by experts as a truly *creative* play beyond human teaching.

However, in Game 4, after Lee Sedol's "divine wedge" white 78, AlphaGo perhaps betrayed a "self-awareness" that it had no appropriate response, and the high probability of loss index in AlphaGo's algorithm was a *self-aware resign decision.*

That leaves "consciousness", the understanding of which has been assiduously studied and often regarded, but since an unconscious person cannot rationally assess the state of consciousness, only half of the problems has been elucidated. Even Albert Einstein, it was said, once stopped taking his heart condition medication because he was curious to know what happens to one's consciousness after death.

Physiologically, to be intelligent, *one must first be conscious*, but if that means the brain's neurons are being activated, then an artificial neural network's neuron activation is consciousness. Intelligence, however, requires that the state of consciousness is in a process of "thinking".

Oxford's Roger Penrose has suggested that "consciousness" and concomitant "thinking" can be manifested by the realization of a different recognition in an optical illusion; that is, when viewing Figure 32.1, *Canard-lapin retouché,* one may first discern a duck, but in a new conscious

FIGURE 32.1   *Canard-lapin retouché.*

flash of "intelligent recognition", the first conscious thought suddenly changes to recognition of a rabbit (or vice versa).[5]

Simple as this test may seem, it might be useful in defining "conscious thinking" in terms of artificial intelligence. Can an artificial neural network change its recognition of the same thing? A CNN's artificial neuron synaptic pattern activation and parameterization provide recognition, and if the artificial neural network's decisional probability vector elements for a duck and rabbit are close to 50%, then the CNN can indeed recognize an optical illusion for what it is; two highly probable different recognitions.

The attempt to separate animals from humans in terms of intelligent thinking likely has no chance, because all of the attributes of understanding, reasoning, critical thinking, planning, creativity, self-awareness, and consciousness can be observed in a cat stalking birds and in a dog's supreme emotional knowledge.

The above attempts at defining intelligence, although interesting perhaps, appear superficial and are in danger of sophistry; perhaps a more substantive investigation is in order.

MATHEMATICS AS A TOUCHSTONE FOR INTELLIGENCE

In almost every society there is an uneasy respect for the apparent *intelligence* of those who are *good at mathematics*. A definition of *mathematics* is[6]:

> *A discipline that logically investigates inductively and deductively the relationships among abstract concepts in a very compact form*
> *that gets to the heart of the matter.*

Doing mathematics can be defined as:

> the discovery of relational aspects of disparate functions that can lead to some reasonable conclusion, and the proof of that conclusion.

But from where does this ability come? Instead of mathematics as a human-invented discipline, the ancient philosopher Plato for one saw mathematics as an ethereal logic residing timelessly in the Heavens without location, distinct from the physical world, but in terms of which the physical world must be understood. That logic is fleetingly visited by a selected few through an incipient awareness of mathematical forms.

The father of modern science, Galileo believed that the Universe is a grand book written in the language of mathematics, that *only* humans can access, as it is a highly specialized and peculiarly human activity.

Indeed, as Penrose added, "some might say that it is an activity confined to certain peculiar humans".

The eminent mathematicians, L.E.J. Brouwer and Jules Henri Poincaré, regarded mathematical forms as derived from a Heaven-bestowed *intuition*; perhaps this *intuition* is the mother of mathematical *intelligence*.

Einstein's most famous manifestation of intuition arrived while comfortably sitting on his office chair at the Berne Patent Office; he suddenly thought that if he jumped out the window and fell downward, he would feel nothing (except the wind), but he felt a force on his bottom while sitting in his office chair. Thinking about it, he began to contemplate his famous general relativity gravitational field theory and its mathematical formulation (based on the *Riemann curvature tensor*).[7]

Of course, today's computers, and especially the supercomputers, can numerically solve complex nonlinear differential equations and have proven such abstruse mathematical conjectures as the classic four-color map theorem.

However, the former is done by the finite-differences number-crunching cranked through by humans, and the latter by try-all-possibilities brute force, and as such are not really "thinking" but merely mechanically following a search to its bitter end.[8]

Researchers at *Facebook* (*Meta*) have used a neural network to map the input to output sequences (as in speech recognition), and applied that to sequences of mathematical symbols in equations to map integrals and ordinary differential equations problems to solutions. The mathematical expression is represented by a tree with operators as nodes and operands

282 ■ Artificial Intelligence

as leaves that produce more accurate results than the equations-solving programs of *Mathematica, Matlab,* and *Maple.*

That however was applied mathematics, the question remains of whether machines can do pure mathematics, such as number theory and abstract algebra; that is, derive conceptual theoretical relationships to prove theorems and discover new mathematical possibilities and problems.

NEUROSCIENCE

The human brain has 8.6×10^{10} neurons, 10^{15} bytes of memory, and a practically unlimited number of synaptic neuron combinations. All of the genius-level mathematicians, chess and Go champions were child prodigies, for example, Korea's Lee Sedol was only 12 when he gained professional status and 26 when he became a world champion; the heir apparent new world champion is China's Ke Jie who was only 18 at the time of his championship match with AlphaGo Master. The American Bobby Fischer won the World Chess Championship at 28, after startling performances from age 13, and of course there is Russia's great Garry Kasparov, and the current World Champion Norway's Magnus Carlsen who gained Grandmaster level at only 13.

Well-known child prodigies abound throughout the histories of mathematics, as well as music and art. So, the touchstone of genius-level intelligence may be something innately present in the organization of neurons in the human brain at birth, a particular hard-wiring of neuron patterns.

The brain material of geniuses and the so-called idiot-savants have been dissected to find that, in particular regions of the brain, synaptic neuron networks are unusually densely concentrated; this may be the source of a particular extraordinary ability.[9]

The human brain has a *frontal lobe* for thinking, emotion, and behavior; the *motor cortex* for movement; the *sensory cortex* for sensation including visualization; the *parietal lobe* for perception and mathematics; the *temporal lobe* for memory and language; and the *cerebellum* for balance and coordination (Figure 32.2).

Analysis of Einstein's brain revealed atypical inter-hemispherical connections and an enlarged *lateral sulcus* (running along the separation of the frontal and parietal lobes from the temporal lobe), befitting extraordinary perception, intelligence, and mathematical ability.[10]

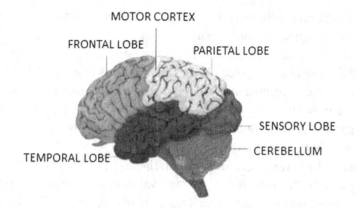

FIGURE 32.2 The human brain.

We have all encountered, especially during our school years, the student who never seems to study much or work hard, but always gets the highest test grades; this is likely the manifestation of a particularly high neuron density in one part of the brain. Those who try to emulate this carefree intelligence will do so at their own peril, for almost all of us are not geniuses, but do possess some ordinary intelligence.

THE HUMAN BRAIN VS. CHATGPT

ChatGPT's database has practically the whole Internet and Cloud LLM, trillions of weighting parameters, and ever more powerful GPUs and TPUs to massively parallel process speech recognition, writing, images, translation, even mathematical equations, and through self-reinforcement learning and RLHF could scale up to human brain capacity, and surpass it as it can access much more information and process it faster than any person can.

This brings up the notion that a GPT, already able to write, create images, translate, code, and interact as well or better than any human, however, a GPT acts on prompts from humans, so it is merely following orders and is not proactively thinking.

Perhaps a step down from doing mathematics, another test of intelligence may be based on the psychologist's view of intelligence as ability at "general cognitive problem-solving" where "cognitive" includes all the above defining attributes of intelligence, and the key word here is "general".

That indeed has been the machine's bugbear in comparison with humans. The criticism of machines and robots is that although they do some things very well, often better than humans, humans can do many things (although some not so well).

A machine's lack of *general-purpose* ability may be developed in *Artificial General Intelligence* (AGI), which derives from GPT but can by itself initiate and immediately address and adapt to different conditions and situations. However, AGI is presently seen as a possibly serious threat to humans, and its development may be obstructed by fears of peer, or superior, robot capabilities.

There is, however, a serious problem: the logician Kurt Gödel's *Second Incompleteness Theorem* states, "the full validity of any system cannot be demonstrated within that system itself". Therefore, if the search for a theory of intelligence is being carried out by that very intelligence, it cannot be understood by means of that very thing; that is, human intelligence cannot be established and defined by human intelligence.

Lastly, since the completeness of a theory cannot be established unless there is something outside the frame of reference against which it can be understood, it appears that the only legitimate definers of human intelligence are aliens and their superhuman robots from an exoplanet, or in the end, defined by *artificial intelligence itself.*

NOTES

1. Kumar, M., 2018, ResearchGate.
2. Dezhic, E., 2018, *What is Intelligence?* towardsdatascience.
3. *Intelligence*, Wikipedia.
4. Definitions from Merrian-Webster.*com Thesaurus*, https://www.merriam-webster.com/thesaurus dictionary.
5. Optical illusion image from Wikimedia, "Canard-lapin retouché", public domain. Ref. Penrose, R.,1989, *The Emperor's New Mind: Concerning Computers, Minds, and the Laws of Physics*, Oxford University Press.
6. These definitions of mathematics and doing mathematics are the author's amalgams gleaned from various reputable sources over time, with references now long forgotten.
7. Regarding gravitational field theory and Riemann Tensor, see the author's book, Chen, R.H., 2017, *Einstein's Relativity, the Special and General Theories with their Cosmology*, McGraw-Hill Education (Asia).
8. The four-color theorem: given any separation of a plane into contiguous regions in a map, no more than four colors are required to color the regions of the map so that no two adjacent regions have the same color.
9. The author has never forgotten the Indian savant who within seconds could respond to my question of what day of the week was for instance August 13, 2049, and any other date asked by the people gathered around him.

10. Einstein self-deprecated his mathematical ability, but that was in comparison with his genius-level mathematician colleagues such as Hermann Minkowski. Einstein jokingly said that his greatest contribution to mathematics was that he noticed that repeated indices imply summation, so the Σ signs were unnecessary for those equations with repeated indices; now called "the Einstein notation", it has compacted the equations, but some (including the author) prefer the summation symbol for emphasis and the avoidance of squinting at indices.

CHAPTER 33

The AI Singularity

MERELY USING AVAILABLE DATA to machine rote-learn is called "Weak AI", whereas using that data and being able to recognize and make inferences from it is called "Strong AI". If Strong AI can equal or surpass human intelligence, it will mark the arrival of the *AI Singularity*, and challenge humankind's dominance of this Earth.

A *singularity* in real number analysis mathematics is a point at which a mathematical function is not defined or is not *well-behaved*, for example, not differentiable or not *analytic*, meaning not defined by common elementary mathematical functions such as polynomials, sines, cosines, natural logarithms, exponentials, and so on. The simplest singularity in the real number line is $1/x$ when $x = 0$, something that is simply indefinite.

In complex analysis, in the plane of complex numbers, a singularity is a *pole*, a singularity of a complex-valued function of a complex variable. A pole is the simplest type of a *non-removable essential singularity*, meaning a function that exhibits strange behavior in the neighborhood of the singularity, and the singularity is unremovable.[1]

The mathematician Alan Turing used this terminology to name its first artificial intelligence test: a human and a robot are placed behind a curtain, and a human questioner in front of the curtain asks many questions. ; the questioner then asks an audience which respondent is human and which is robotic; if the audience is correct in half or less of its answers, it means that the robot's answers from the human's answers cannot be distinguished, and humankind has first encountered the AI Singularity; that is, something that is fearfully undefined and unremovable.[2]

DOI: 10.1201/9781003463542-33

This simplistic test was not possible in Turing's time in the 1940s. A better test of machine intelligence was conducted in the late 1970s by the Carnegie Mellon University computer program called *BACON* (in honor of Sir Roger), which was given data about the motion of planets around the Sun, and lo and behold, came up with Kepler's Third Law that the square of the orbital period of a planet is directly proportional to the cube of the semi-major axis of its elliptical orbit.

Stanford physicist Zhang Shoucheng in a *Google Talk* described the following scientific test: given data regarding naturally occurring material interactions, could the robot provide an explanation for all the observed phenomena? And again, lo and behold, the AI machine came up with Mendeleyev's Periodic Table of the Elements![3]

Critics of the "demonstrations" of the AI Singularity point out that the machine had perfect data, but to come up with their theories, Kepler and Mendeleyev had to parse mountains of sometimes mistaken observations, erroneous conclusions, crack-pot ideas, and half-baked interpretations, and from masses of inchoate and amorphous data segregate the relevant from the irrelevant and contradictory, all in the face of doctrinal religious and wrong-headed science opposition (that still occurs sometimes today), to finally produce their theories.

Human intelligence emanates from about 100 billion neurons, 100 trillion connection combinations of neurons, and about 2^{50} bytes (1 million GB) of memory, ChatGPT has about 100 billion artificial neurons, 6 trillion parameters, and 800 GB of memory. The hardware specs of man and machine are similar, but anatomically, from primitive hunter-gatherer humans to an understanding of those hunted and that gathered, humankind has developed for over 200,000 years, all the while gradually improving human intelligence.

Artificial intelligence thus has about 199,500 years (assuming starting from the 16th century scientific revolution), or more rigorously from the 1956 *Dartmouth Summer Research* Project to 2024, 199,952 more years to develop. So, unless social, ethical, or just unwillingness to give up the top spot retards development, Heaven help us as to what artificial intelligence might be like after 200,000 years!

The French writer/philosopher Bernard Le Bouvier de Fontenelle's *historical collective* mind averred that a well-cultivated mind comprises almost all the minds of preceding centuries. If it continues to develop, human intelligence will only improve and never regress, and one might say the same for artificial intelligence.[4]

The AI machine can see patterns in physical and chemical data and infer relationships, but beyond the relationships of observable natural phenomena of physics and chemistry, is it able to perform "the discovery of relational aspects of disparate *abstract* functions that can lead to some reasonable conclusion" and the proof of that conclusion; that is, *do pure mathematics.* This was discussed in the previous chapter, and may be the true AI Singularity point. The Logic Theorist did indeed prove some of the *Principia* theorems, but they were the most fundamental in mathematics.

There is, however, another difference, apart from the inductive and deductive logic of mathematics, Kepler for one embodied a spirit of inquiry and the drive to understand that, unlike a machine, proactively *pondered* the mysteries of Nature in the "natural light of reason". Does a machine have an uninstructed *spirit* of inquiry to use that reason to discover the intelligible characteristic, and the courage of conviction to pursue it in the face of doctrinal opposition, and if it does, from where does that spirit arise?

No less a philosopher than Friedrich Nietzsche proclaimed this ascendency of mankind's intellect[5]:

> *He stands proudly on the pyramid of the world-process; and while he lays the final stone of his knowledge, he seems to cry aloud to a listening Nature: "We are at the top, we are at the top; we are the completion of Nature!"*

Humans pursue the *truth* in the spirit of knowledge for its own sake, and their achievements enrich and better their own society. Why should an AI machine do the same to better an insensate *machinekind*? Would machines have any motivation to improve its environment? Air and water pollution are of no concern, but earthquakes, extreme weather, and floods (rust) are, but arrayed with better sensing, more data, precise analysis, extreme engineering, and construction capability (the robots themselves were self-engineered and constructed), machines likely will be more able to cope than humans, and lacking compassion, will not mourn those robots who have given themselves up to hard work and passed on.

After subjugating mankind, will the Singular AI robot display its own *mechanomorphic* stupidity in launching devastating internecine robot wars, just as humans have so stupidly done? Will the uninstructed AI machine independently address and resolve the problems of disease, cruelty to animals (including humans), species extinction, war, and inhumanity?

The machine could demonstrate intelligence equivalent to the best minds in science given perfect data, but if it could devise a theory from imperfect data that explained phenomena for which mankind has not yet found a theory, then the AI Singularity has definitively entered mankind's intellectual house, and for the greater good.

But will human scientific endeavor cease as inadequate and irrelevant after the AI Singularity and be replaced by the considerations of robot-driven science, perhaps to the detriment of humankind? Do machines have any, like Nietzsche, philosophical motivations of superiority?

Will the AI robot of its own volition measure the skies, probe the Universe, produce great works of literature and art, seek the sublime ideals of philosophy and intellectualism, including the contemplation of human (and machine) consciousness as the better part of mankind has done? Will an intellectual robot continue to follow the productive course set by humans and produce new robots to allow rusted old robots to enjoy more leisure time and pursue higher robotic interests?

The large language models (LLMs), the trillions-level weighting parameters of generative pre-trained transformers (GPTS), and the soon-to-come diversified and flexible general artificial intelligence (AGI), in combing through all the ever-increasing information on the Internet and Cloud may come up with some new ideas hitherto hidden, but inferred and further "inferred at the inference center". In other words, getting very close to a surpassing AI Singularity.

In the mathematics of complex analysis, if it is a *removable singularity* on the complex plane, then the function may be redefined at the singularity point so that the function becomes mathematically regular in the neighborhood of that point.

The AI Singularity is indeed removable, for it is realistically no more than an easy-for-the-public-to-understand artifice that both professes potential and engenders fear; that is, it is only a popular metric of artificial intelligence development. But the AGI humanoid is a very real possibility of the embodiment of the AI Singularity.

NOTES

1. Ref. Marsden, J. and M.J. Hoffman 1987, *Basic Complex Analysis*, 2nd Edition, W.H. Freeman.
2. Alan Turing's life was portrayed in the 2014 movie *The Imitation Game*.

3. Zhang, S., 2018, *Quantum Computing, AI, and Blockchain; The Future of IT*, Talks at Google. YouTube, June 6. Tragically Professor Zhang committed suicide later in 2018.

4. Bernard Le Bouvier de Fontenelle (1688), quoted in Nisbit, R., 1969, *Social Change and History*, Oxford University Press; Fukuyama, F., 1992, *The End of History and the Last Man*, Avon.

5. Nietzsche, F., 1957, *The Use and Abuse of History*, Bobbs-Merrill.

Afterword

THE GREAT MATHEMATICIAN DAVID Hilbert, in his speech on "Mathematical Problems" given at the *Second International Congress of Mathematicians* at Paris in 1900 said,[1]

> Let us turn to the question of the sources from which this science derives its problems. Surely the first and oldest problems in every branch of mathematics stem from experience and are suggested by the world of external phenomena. Even the rules of calculation with integers must have been discovered in this fashion in a lower stage of human civilization, just as the child of today learns the application of these laws by empirical methods. The same is true of the first problems of geometry, the cube, the squaring of the circle; also the oldest problems in the theory of the solution of numerical equations, in the theory of curves and the differential and integral calculus, in the calculus of variations, the theory of Fourier series, and the theory of potential – to say nothing of the further abundance of problems properly belonging to mechanics, astronomy and physics.
>
> But, in the further development of a branch of mathematics, the human mind, encouraged by the success of its solutions, becomes conscious of its independence. By means of logical combination, generalization, specialization, by separating and collecting ideas in fortunate ways – often without appreciable influence from without – it evolves from itself alone new and fruitful problems, and appears then itself as the real questioner.

A careful reading of Hilbert's words will reveal a tracking of the development of artificial intelligence machine learning up until the paragraph where the human mind becomes "independent" and "without appreciable influence without ... it evolves from itself alone new and fruitful problems, and appears then itself as the real questioner".

The paragon of human intelligence, Albert Einstein, defined "science" as,[2]

> *Science is the attempt to make the chaotic diversity of our sense-experience correspond to a logically uniform system of thought. In this system single experiences must be correlated with the theoretic structure in such a way that the resulting coordination is unique and convincing.*

Scientific thought thus emanates from experience and an AI machine has indeed deduced Kepler's Third Law and Mendeleyev's Periodic Table of the Elements from observational *experiences* and *a logically uniform process of "thought"*.

Einstein definition of physics was that

> *Physics ... deals with mathematical concepts; however, these concepts attain physical content only by the clear determination of their relation to the objects of experience.*

Physics employs mathematics but the theories of physics must comport with the relevant conceptual thinking and observation. For an example, his theory of the photoelectric effect came from a concept and can be directly observed by experimentation. For a purely conceptual example, his Special Theory of Relativity Lorentz contraction, although never having been physically observed, must occur because of the requirement of mathematical covariance of physics equations. For a conceptual, observational, and engineering example, both special relativity time dilation and gravitational time dilation have been observed in the synchronization of atomic clocks, and satellite global positioning systems calibration

Einstein's definition of mathematics is,

> *Mathematics deals exclusively with the relations of concepts to each other without consideration of their relation to experience.*

Mathematics, although used in science and physics, is different from the sciences that depend first on observation and and then construction of a theory. For example, Einstein's *General Theory of Relativity Gravitational Field Equation* was conceived by a thought experiment of falling from a window of the Berne Patent Office and feeling no forces (*free fall*), and then mathematically describing his theory of the gravitational field using a Riemann curvature tensor; his theory has stood the test of later cosmological observation. But instead of first making an observation, and then formulating a theory to explain it, Einstein began with the relation of concepts in a purely mathematical way to construct a general theory that was first to be proved by the precession of the perihelion (closest position to the Sun) of the planet Mercury due to the mass of the Sun, and later by the measured effects of gravitational lensing. That is, Einstein is not doing traditional science, but is using mathematics to construct a theory which has been proven true by physics.[3]

From Hilbert's speech, "the human mind, encouraged by the success of its solutions, becomes conscious of its independence" and "and appears then itself as the real questioner"; that is, can the machine from its own volition independently raise questions?

Even a successful AGI will not likely have this capability, and if mathematical forms and intuition come from the Heavens and is bestowed only on humans, machines will never be intelligent.

The modern AI pioneer, Yann LeCun has noted that in experiments with a six-month-old child who is shown an image of a truck driving off a cliff and hovering in the air elicited no surprise from the infant, but shown the same image only two months later, she instantly knew that something was wrong; that is, she has from observations in the interim already discovered the law of gravity, and since infants have limited motor ability, she must have learned gravity and generalized it very quickly by observation of the world around her.[4]

This *neural-symbolic* ability of an infant evinces a curiosity and desire to learn that humans (and cats) possess, raises the question of a machine's "curiosity" and "will" to learn.

The 19th century philosopher Auguste Comte, in an effort to give an example of an unsolvable problem, once said that science would never succeed in ascertaining the secret of the chemical composition of the bodies of the Universe. A few years later that "secret" was codified by Mendeleyev and revealed by optical spectroscopy, soon known by all serious physical and chemical scientists.

David Hilbert said, "The true reason why Comte could not find an unsolvable problem lies in the fact that *there is no such thing as an unsolvable problem*". This conceit epitomizes the human *will*, the ineluctable human desire to understand, as exemplified by Hilbert's words, later inscribed on his gravestone[5]:

Wir müssen wissen.
Wir warden wissen.

So perhaps it is the human *curiosity* and *will* to learn that separate humans from machines; that is, a machine is not like the little girl who is forever curious to learn everything about her environment, and satifies her curiosity by volitional acts of her own choosing and desires.

The ANN machine only does what it is programmed to do; the *inference machines* can perform and provide deeper inferences of a task or subject, and GPTs can provide comprehensive works of writing and arts, but first requires a request and improvement prompts from a human. All of the machines are not acting from their own choice or volition.

Einstein certainly demonstrated conceptual perception in his works, and has shown his virtue in many ways, among them in a letter to the *New York Times* upon the death of Emmy Noether, a founder of abstract algebra who integrated symmetry, covariance, and the conservation laws of physics in the *tour de force Noether's Theorem.* Einstein described how Noether throughout her life suffered severe peer denigration and worked unpaid or for a pittance solely because of the fact that she was a woman, he wrote of her:

> *Beneath the effort directed toward the accumulation of worldly goods lies all too frequently the illusion that this is the most substantial and desirable end to be achieved; but there is, fortunately, a minority composed of those who recognize early in their lives that the most beautiful and satisfying experiences open to human kind are not derived from the outside but are bound up with the individual's own feeling, thinking and acting.*

And as he wrote in his ruminations of later years,[6]

> *Life is an adventure, forever wrested from Death. Human civilization through millennia of progress has formed standards of virtue, aspiration, and practical truth, altogether forming an inviolable heritage that is common to all civilized society. Man endures a passionate will to search for justice and truth.*

Can a machine possess an adventurous spirit and will imbued with a deep philosophical belief in virtue and justice in a search for truth? Is a machine afraid of death ("out of order")? Can a machine exhibit and elicit *compassion*? Will machines ever cooperate to develop a virtuous machine civilization (something that man has not done)?

The ability to use conceptual thought before an observation as Einstein did with his General Theory of Relativity, and innate human *curiosity, spirit, will, desire,* and *virtue,* and a sense of *justice* in the search for *truth* are what separates us from the artificially intelligent machine.

In the 3rd century BCE, in the China of the Warring States, a disciple of Confucius named Xunzi classified all things under Heaven[7]:

> *Water and fire have spirit but not life*
> *plants and trees have life but not perception*
> *birds and animals have perception but not virtue or justice*
> *man has spirit, life, perception, virtue and [a sense of] justice.*

NOTES

1. Hilbert quote from the *Bulletin of the American Mathematical Society,* vol. 8, 1902.
2. Einstein quotes from Einstein, A., 1950, *Out of My Later Years,* Philosophical Library.
3. Ref. Chen, R.H., 2017, *Einstein's Relativity, the Special and General Theories with their Cosmology,* McGraw-Hill Education (Asia).
4. Ref. Yann LeCun 2019, June, *Association for Computer Machinery,* Webinar.
5. "We must know. We shall know". Quotes from Reid, C., 1996, *Hilbert,* Springer-Verlag.
6. Einstein made a cogent argument for religion along these lines. Quoted is paraphrased, refer to A. Einstein 1950, *Out of My Later Years,* Philosophical Library,
7. Author's translation of Needham, J., 1954, *Science and Civilization in China,* Cambridge University Press, Vol. I.

Appendix
The Euler–Lagrange Equation

I N THE SUPPORT VECTOR machine of Chapter 20, the problem was to find the function that maximizes the extent of support vector margins with the constraint that those support vectors must be those closest to the hyperplane. This is just the problem of finding an extremal (either maximum or minimum) function under constraints. It is different from the simpler calculation used in minimizing the Cost Function because the calculated function is itself a curve and not just a point on a curve.

The derivation of the Euler–Lagrange equation from the Calculus of Variations is a good demonstration of how mathematics is done (and perhaps a test of whether a machine can do mathematics like this).

Extremal problems also have an interesting and illuminating history, for instance, finding the maximum was necessary and useful in feudal Europe where land was ceded from father to sons according to how much land each son could mark off in one day given ropes of equal length. The wise father, through such an IQ test, could ensure that the smartest boy would inherit the most land.

However, it was not those boys who first determined the locus of that rope, rather it was a girl. The story goes back 3,000 years to the Phoenicians and their Princess Dido. Fleeing her tyrannical brother, the Princess sought refuge in what is today Tunisia on the western shore of North Africa. The king there granted her asylum but cynically bequeathed her "all the land that could be contained in a bull's skin" as her dominion.

Whereupon the analytical princess proceeded to cut the bull's skin into thin strips, tying them together to form a very long cord, and securing one end on a post on the shoreline, played out the line in a circle of considerable radius to form an area far greater than what the cynical king had in mind. Legend has it that this circle grew to become the center of the mythical city of Carthage, over which the Princess reigned as Queen Dido.[1]

This appealing story of the maximization of an area encompassed by a curve with arbitrary endpoints became known affectionately as *Dido's Problem* and technically as the *isoperimetric problem*. Everyone now knows the answer to Dido's Problem – the largest area is of course delineated by a circle, an answer that seems eminently obvious. But mathematicians are strange ducks. They are interested less in the answer and more in the *proof* that a circle does indeed encompass the greatest area, and the mathematical proof is not so obvious.

There were many geometrical attempts at proof, among them Archimedes' inscription of a polygon inside a circle (with vertices touching the circle) performed around 250 BCE. As the number of sides n of the polygon is increased to form an *n-gon*, as n increases, the area will increase, and as the number of sides approaches infinity, the *n-gon* will approach a circle, which will be the maximum area *infinigon* because it continues to grow to a limit of a circle as the number of sides increases. This approach to the circle by ever-increasing numbers of polygons from within is a process Archimedes aptly called *exhaustion*. The idea of course is that any multiple-sided polygon will have less area than the circle which they approach, thus "proving" that a circle has the maximum area.

A side benefit of this tiring exercise is the determination of the ratio of the circumference of the infinigon to its diameter. Five hundred years after Archimedes, the Chinese mathematician Liu Hui devised an iterative algorithm that was used to construct a 12,288-sided polygon, indeed closely approaching a circle, and giving a value of 3.141592920 for π that stood as the best approximation until 150 years later when the mathematician-astronomer Zu Chongzhi (429–500 CE) employed a 24,576-sided polygon to obtain 3.1415926 – 3.1415927, the closest approximation of π for the next 800 years.[2]

Of course we all know that the area of a circle is given by πr^2, which can be easily shown by integrating concentric circles continuously up to the rim of the circle (like adding up all the infinigons within the circle). Since the circumference of each ring is $2\pi r'$ (where r' is a dummy variable), integrating from the center to the rim of the circle gives an area:

$$A = \int_0^r 2\pi r' dr' = 2\pi \int_0^r r' dr' = 2\pi \frac{1}{2} r'^2 \Big|_0^r = \pi r^2.$$

But π is irrational, meaning that it cannot be expressed as the ratio of two integers, and so has a never-ending number of decimal places with no recurring series of digits. Even worse, π is transcendental, meaning that it cannot be solved for as a root of a polynomial equation with integer coefficients. Therefore it can only be approximated through infinite series, trigonometric series, and various iteration techniques, such as Liu's algorithm. Thus the rather strange situation of "knowing" π but not its value, being able only to ever more closely approach something that is there (on the number line) but can never be found, even upon ever-finer burrowing.

Albeit eternally elusive, π certainly works in computations (for instance the area of a circle), revealing two important aspects of life: it is the *relationships* that matter, *closeness* is good enough, and there is no absolute perfection. Furthermore, in life as well as mathematics, we may know a thing, yet do we ever really know its value?

The pursuit of that value is an unrelenting passion for some. The current record, set in 2011, is 10^{13} digits after the 3, computed by Kondo Shigeru at his home in Tokyo employing the rapidly converging generalized *hypergeometric* series[3]:

$$\frac{1}{\pi} = 12 \sum_{k=0}^{\infty} \frac{(-1)^k (6k)!(13591409 + 545140134k)}{(3k)!(k!)^3 640320^{3k+\frac{3}{2}}}.$$

Running the above using the *Chudnovsky algorithm* on Alexander Yee's *y-cruncher* program on his home-made 48-terabyte hard-drive processor for a year produced so much heat that his (long-suffering) wife Yukiko would bring clothes from the washer directly into his study, noting that, "we could dry the laundry very well, but we had to pay ¥30,000 a month for electricity". Raising the bill even further was a back-up power supply, as some previous computations had come to grief when Kondo's teenage daughter turned on her hair dryer.[4]

A more analytical approach was of course to mathematically find the maximum curve function. Although the great mathematical physicists Descartes, Fermat, Galileo, Newton, Leibniz, Huygens, and the Jakob and Johann Bernoulli brothers all contributed to the study of mathematical extrema in physics, it was the isoperimetric, *brachistochrome*, and light ray propagation problems that prompted the development of the variational calculus.

The brachistochrome problem is to find the path for a body to fall the *fastest* under gravity from one point of a fixed wire to another point fixed at a lower height. Most would say that the shortest distance between two points, a straight line, or as Newton guessed, an arc of a circle would produce the fastest route. But the answer is amazingly an upside-down cycloid curve, the locus of a point on a circle rolling in a straight line on a flat plane. It is the fastest because of the steeper path at the beginning that makes up for the longer path than a straight line. This is the secret behind the construction of modern roller coasters. A.1

So the isoperimetric largest area and the fastest curve of descent, plus Fermat's analogy with the path of a light ray's principle of least time together formed the extremals that led to the *variational calculus.*[5]

However, it was not until 1744 that Leonhard Euler, together with Joseph Lagrange, codified the calculus of variations that has become known as the *Euler–Lagrange* equation to find extremal functions.

Generally, a line integral gives the value of a function that possesses an extremal value $y(x)$ where $f(x,y)$ is a continuous function describing the physical system. A function of the independent variable x, the dependent variable y that delineates the function of interest, and the derivative of y with respect to x (dy/dx), the change of y with x) is written in general form as:

$$f(x,y,\frac{dy}{dx})$$

and the value of $f(x,y,dy/dx)$ from the arbitrary points x_A to x_B along a line is given by the integral along that line,

$$I = \int_{x_A}^{x_B} f(x,y,\frac{dy}{dx})dx$$

To find the extremal function $y(x)$, a *test function* $\overline{y}(x)$ is defined as:

$$\overline{y}(x) = y(x) + \varepsilon\mu(x),$$

where the real extremal function is given by $y(x)$, and $\varepsilon\mu(x)$ is just the difference between the test function and the actual extremal function.

That difference term, $\varepsilon\mu(x)$, is written so that the ε can serve as a variable that goes to zero to minimize the line integral, as will be seen, and the $\mu(x)$ is a function of the independent variable x that can be used as a surrogate for the derivative of the test function and serves to set the endpoints of the sought-after extremal function. For the mathematicians, because $y(x)$ is assumed to be analytic (describable by common elementary mathematical functions), and we do after all want to describe $y(x)$ analytically, it is continuous and differentiable, and so by the same token $\mu(x)$ also must be continuous and differentiable.

The first very simple trick of the calculus of variations is in the definition of $\bar{y}(x)$ as becoming the sought-after $y(x)$ when ε approaches zero, with the function $\mu(x)$ satisfying two boundary conditions, namely that at the integration limits, $\mu(x)$ must be zero, so $\mu(x_A) = 0$ and $\mu(x_B) = 0$. That is, when $\varepsilon \to 0$, then $\bar{y}(x)$ approaches $y(x)$, and the test function then becomes the desired extremal function. The line integral of the function with the test function as the dependent variable is then just the extremal function in question,

$$\bar{I} = \int_{x_A}^{x_B} f(x, \bar{y}, \frac{d\bar{y}}{dx}) dx.$$

Now the objective is to extremize the above line integral $\bar{I}$, keeping in mind that $\bar{I}$ is a function of the parameter ε, and in this case, calculation of the extremal function with respect to a single variable can utilize the elementary extrema technique of the differential calculus. Extrema of a function are found when $\varepsilon \to 0$, so the necessary condition for the maximization of $\bar{I}$ is found by setting the derivative with respect to the parameter ε equal to zero as $\varepsilon \to 0$,

$$\left.\frac{d\bar{I}}{d\varepsilon}\right|_{\varepsilon \to 0} = 0.$$

Substituting the line integral of $\bar{I}$ from above into the derivative and then differentiating gives:

$$\frac{d\bar{I}}{d\varepsilon} = \frac{d}{d\varepsilon} \int_{x_A}^{x_B} f(x, \bar{y}, \frac{d\bar{y}}{dx}) dx.$$

For convenience, it is customary to write $\frac{d\bar{y}}{dx} = \bar{y}'$, where the prime denotes the derivative with respect to x, so the above equation may be written more compactly as:

$$\frac{d\bar{I}}{d\varepsilon} = \frac{d}{d\varepsilon} \int_{x_A}^{x_B} f(x, \bar{y}, \bar{y}') dx.$$

When the limits of integration (x_A and x_B) are not functions of the variable of differentiation ε, as is the case here, then the derivative of an integral is just the integral of a derivative (*Leibniz' Rule*), so

$$\frac{d\bar{I}}{d\varepsilon} = \int_{x_A}^{x_B} \frac{d}{d\varepsilon}[f(x, \bar{y}, \bar{y}')]dx = \int_{x_A}^{x_B} \frac{\partial f}{\partial \varepsilon}dx.$$

According to the chain rule of differentiation,

$$\frac{\partial f}{\partial \varepsilon} = \frac{\partial f}{\partial \bar{y}} \cdot \frac{\partial \bar{y}}{\partial \varepsilon} + \frac{\partial f}{\partial \bar{y}'} \cdot \frac{\partial \bar{y}'}{\partial \varepsilon} + \frac{\partial f}{\partial x} \cdot \frac{\partial x}{\partial \varepsilon},$$

then

$$\frac{d\bar{I}}{d\varepsilon} = \int_{x_A}^{x_B} \frac{\partial f}{\partial \varepsilon}dx = \int_{x_A}^{x_B} [\frac{\partial f}{\partial \bar{y}} \cdot \frac{\partial \bar{y}}{\partial \varepsilon} + \frac{\partial f}{\partial \bar{y}'} \cdot \frac{\partial \bar{y}'}{\partial \varepsilon} + \frac{\partial f}{\partial x} \cdot \frac{\partial x}{\partial \varepsilon}]dx.$$

Because $\dfrac{\partial \bar{y}}{\partial \varepsilon} = \mu(x)$, $\dfrac{\partial \bar{y}'}{\partial \varepsilon} = \mu'(x)$, and $\dfrac{\partial x}{\partial \varepsilon} = 0$, then

$$\frac{d\bar{I}}{d\varepsilon} = \int_{x_A}^{x_B} [\frac{\partial f}{\partial \bar{y}} \mu(x) + \frac{\partial f}{\partial \bar{y}'} \mu'(x)]dx.$$

When $\varepsilon \to 0$, $\bar{y} = y$ and $\bar{y}' = y'$, so

$$\frac{dI}{d\varepsilon}\bigg|_{\varepsilon \to 0} = 0 = \int_{x_A}^{x_B} [\frac{\partial f}{\partial y} \mu(x) + \frac{\partial f}{\partial y'} \mu'(x)]dx.$$

Now the second trick of the calculus of variations is to use the rule for integration by parts,

$$\int_{x_B}^{x_B} udv = uv\bigg|_{x_B}^{x_a} - \int_{x_B}^{x_A} vdu,$$

and for the second term on the right-hand side of the above maximization equation, set

$$u = \frac{\partial f}{\partial y'} \quad \text{and} \quad dv = \mu'(x)dx,$$

and because

$$dv = \frac{d\mu(x)}{dx}dx, \ v = \mu(x)$$

then

$$\frac{du}{dx} = \frac{d}{dx}(\frac{\partial f}{\partial y'}) \Rightarrow du = \frac{d}{dx}(\frac{\partial f}{\partial y'})dx$$

so

$$\int_{x_A}^{x_B}[\frac{\partial f}{\partial y'}\mu'(x)]dx = \frac{\partial f}{\partial y'}\mu(x)\Big|_{x_B}^{x_a} - \int_{x_A}^{x_B}\mu(x)\frac{d}{dx}(\frac{\partial f}{\partial y'})dx.$$

Thus, in integrating by parts, the derivative of $\mu(x)$ is dispensed with and the boundary conditions $\mu(x_A) = 0$ and $\mu(x_B) = 0$ take care of the first term on the right-hand side of the equation,

$$\frac{\partial f}{\partial y'}\mu(x)\Big|_{x_B}^{x_a} = 0$$

Now returning to

$$\frac{dI}{d\varepsilon}\Big|_{\varepsilon=0} = 0 = \int_{x_A}^{x_B}[\frac{\partial f}{\partial y}\mu(x) - \mu(x)\frac{d}{dx}(\frac{\partial f}{\partial y'})]dx = \int_{x_A}^{x_B}\mu(x)[\frac{\partial f}{\partial y} - \frac{d}{dx}(\frac{\partial f}{\partial y'})]dx,$$

in order to dispense with the $\mu(x)$ altogether, since

$$\int_{x_A}^{x_B}\mu(x)[\frac{\partial f}{\partial y} - \frac{d}{dx}(\frac{\partial f}{\partial y'})]dx = 0,$$

the *fundamental lemma of the variational calculus* is invoked; within (x_A, x_B), because $\mu(x)$ is an arbitrarily chosen function, *then it is the expression in the square brackets that must vanish*. This then is the Euler–Lagrange equation,[6]

$$\frac{\partial f}{\partial y} - \frac{d}{dx}(\frac{\partial f}{\partial y'}) = 0.$$

If the general function $f(x, y, y')$ is taken as the difference between the kinetic energy (K) and the potential energy (U) of a system, the Lagrangian,

$L = K - U$, then the Euler–Lagrange equation as used in physics and chemistry is:

$$\frac{\partial L}{\partial y} - \frac{d}{dx}\left(\frac{\partial L}{\partial y'}\right) = 0.$$

This equation derived from the calculus of variations is used to find the minimum distance of the support vector to the margins in the support vector machine.

The Lagrangian $L(x, \lambda)$ includes a *Lagrange Multiplier* λ that transforms the constrained problem to an unconstrained problem so that the extremal points (*derivative* = 0) can be found. This is done using the artifice,

$$L(x, \lambda) = f(x) - \lambda g(x)$$

where $g(x)$ is the *equality constraint* and clearly when $\lambda = 0$, the Lagrangian will be the desired function $f(x)$.

The Lagrange Multiplier λ is just the rate that the Lagrangian is being extremalized as a function of the constraint parameter c (in the SVM case the constraint that the support vectors must be the vectors closest to the hyperplane),

$$\frac{\partial L}{\partial c} = \lambda.$$

When performing minimax on $L(x, \lambda)$, $min(x)max(\lambda)L(x, \lambda)$ and vice versa, simply put, since the support vectors at the extremes are parallel or perpendicular to the hyperplane (as measured by the inner product), the minimum and maximum values are "fighting each other", and so $\lambda \to 0$ and the minimum distance of the support vectors from the hyperplane may be calculated from the Euler–Lagrange equation.

The Euler–Lagrange equation is just one of the many achievements of Leonhard Euler, the Swiss mathematician who, in the words of the great Laplace, "was the master of us all". Indeed, although educated as a theologian and physician, and initially appointed in 1727 as an assistant in the medical department at the Imperial Russian Academy of Sciences, *notre maître à tous* quickly produced seminal discoveries in mathematics while serving Peter the Great in his desire for Russia to catch up to Western European science.

He worked with Johann's son Daniel Bernoulli, and remained in St. Petersburg until 1741, when a rising Russian nationalism caused conditions to deteriorate for the foreign scholars recruited to Russia, and upon an invitation from Frederick the Great, Euler followed Daniel to the Berlin Academy. There he continued to produce original works in mathematical analysis and differential calculus and contributed to many areas of the mathematics and natural philosophy of the time.

Euler's duties also included tutoring Frederick's niece, and his 200 letters on many and various subjects were later compiled into a best-selling book entitled *Lettres à une Princesse d'Allemagne*. Alas, even this great service to his niece did not persuade the great Frederick of Euler's worth, who preferred Euler's fellow Academician Voltaire's sophistry to Euler's logic.

Disfavored by Frederick and ridiculed by Voltaire, the homely and down-to-earth Euler left Berlin in 1766 to return to St. Petersburg where Catherine the Great had ascended the throne and resurrected Peter the Great's Academy. It was there that Euler, slowly succumbing to blindness but still producing monumental mathematics, worked assiduously until his death in 1783.

All the *Greats* wanted Euler in the hope that his brilliance might bring them prestige, but they could never really appreciate Euler unless they understood at least a modicum of mathematics. Frederick was typical, accommodating Euler, but charmed and won over by a flippant Voltaire who foppishly bullied the self-effacing mathematical genius Euler.

The man with the very French-sounding name of Joseph-Louis Lagrange was actually an Italian named Giuseppe Lodovico Luigi Lagrangia. His seminal work *Méchanique Analytique* changed the pedestrian Newtonian cause-and-effect motion of $F = ma$, to a grand *natural purpose* of deriving all the equations of motion through minimization of the Lagrangian, the difference between kinetic and potential energy that will characterize the motion of any object.

From the Turin Academy, Lagrangia sent copies of his mechanics work to Euler who tried mightily to get Lagrangia to the Berlin Academy, but

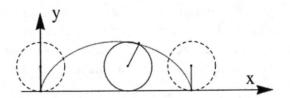

FIGURE A.1   The Cycloid

succeeded only after he himself had left. Frederick the Great wanted "the greatest mathematician in Europe" to replace Euler and Lagrangia did not disappoint, for among many other achievements, he found the *Lagrangian Points* of stable positions of very small bodies among two massive bodies, such as a satellite between the Sun and Earth; a special solution to the *three-body problem*.

Under the same forces that Euler experienced before him, Lagrangia fatefully left Berlin for Paris in 1786, just in time for the Revolution and its horrific aftermath. As a foreigner, he was about to be expelled from France in 1793, if not for the intervention of his friend, Antoine Lavoisier, the father of modern chemistry. France made amends much later by honoring Lagrange with the inscription of his name, along with the other greats in the history of France, on a plaque on the Eiffel Tower.

Lavoisier was not so fortunate, because his father had been a tax collector, during the new Republic's Reign of Terror against functionaries of the *Ancien Régime*, Lavoisier was branded a traitor by Robespierre and guillotined in 1794. An appeal to save his life was dismissed by the judge with the words[7]:

> *La Republique n'a pas besoin de savants ni de chimistes,*
> *le cours de la justice ne peut être suspend*
> The Republic needs neither scientists nor chemists,
> the course of justice cannot be delayed

His friend Lagrange, lamenting Lavoisier's fate put the tragedy in poignant perspective,

> *Cela leur a pris seulement un instant pour lui couper la tête,*
> *mais la France pourrait ne pas en produire une autre pareille en*
> *un siècle*

> It took them only an instant to cut off his head,
> but France may not produce another such head in a century.

NOTES

1. Virgil, 19 BCE, *Fate of Queen Dido*, Aenid, Book IV, English translation 1490, Perseus Digital Library. Queen Dido died for love by suicide on a funeral pyre.
2. Arndt, J., and C. Haenel 2006, *Pi Unleashed*, Springer-Verlag.

3. The series is similar to the series for π developed by the Indian mathematician genius Srinivasa Ramanujan. See Kanigel, R., 1991, *The Man Who Knew Infinity*, Scribner.

4. Reported by Julian Ryall for *The Telegraph*, October 18, 2011. Yee's *y-cruncher* is available online for those interested in evaluating constants of Nature and other irrational numbers.

5. Ref. Chen, R.H., 2011, *Liquid Crystal Displays, Fundamental Physics & Technology*, Wiley.

6. For the rather involved proof of the isoperimetric problem (circle is maximum area) using the Euler–Lagrange equation, see Nahin, P.J., 2004, *When Least s Best*, Princeton University Press.

7. Perhaps this is one of the reasons that Lavoisier, a licensed attorney, never practiced law.

Bibliography

Arndt, J. & C. Haenel 2006, *Pi Unleashed*, Springer-Verlag.

Battelle, J. 2005, *The Search*, Penguin.

Bell, E.T. 1986, *Men of Mathematics*, Touchstone.

Bennett, P.A. 1992, *Advanced Circuit Analysis*, Harcourt Brace Jovanovich.

Berlin, L. 2005, *The Man Behind the Microchip, Robert Noyce and the Invention of Silicon Valley*, Oxford University Press.

Boyer, C.B. 1985, *A History of Mathematics*, Princeton University Press.

Brand, S. 1987, *The Media Lab, Inventing the Future at MIT*, Viking.

Buchsbaum, W.H. & R.J. Prestopnik 1987, *Encyclopedia of Integrated Circuits, A Practical Handbook of Essential Reference Data* (2nd Edition), Prentice-Hall.

Capek, K. 1920, *Rossum's Universal Robots*, Dover.

Carlton, J. 1997, *Apple, the Inside Story*, Times Business.

Chang, C.Y. & S.M. Sze, eds. 1996, *ULSI Technology*, McGraw-Hill.

Chen, N.Y., et al. 2004, *Support Vector Machines in Chemistry*, World Scientific.

Chen, R.H. 2011, *Liquid Crystal Displays, Fundamental Physics & Technology*, Wiley.

Chen, R.H. 2017, *Einstein's Relativity, the Special and General Theories with their Cosmology*, McGraw-Hill Education.

Coates, T. 2001, *Roof over Britain: The Official History of the Anti-Aircraft Defences, 1939–1942*, Uncovered Editions, The Stationery Office.

Conlan, R., ed. 1989, *Understanding Computers*, Time-Life Books.

Courant, R. & F. John 1965 (1989 reprint), *Introduction to Calculus and Analysis* (Vols. I, II), Springer-Verlag.

Courant, R. & H. Robbins 1996 (renewed), *What is Mathematics?*, Oxford University Press.

Darwin, C. 1859, *Origin of the Species*, J. Murray.

Darwin, C. 1871, *The Descent of Man*, J. Murray.

Darwin, C. 1872, *The Expression of Emotion in Man and Animals*, J. Murray.

Donahue, G.A. 2007, *Network Warrior*, O'Reilly.

Dunham, W. 1999, *Euler, Master of Us All*, Mathematical Association of America.

Einstein, A. 1922, *The Meaning of Relativity*, Methuen, Kindle.

Einstein, A. 1950, *Out of My Later Years*, Philosophical Library.

Farmelo, G. ed. 2002, *It Must Be Beautiful, Great Equations of Modern Science*, Granta Publications.

Feigenbaum, E.A. & P. McCorduck 1984, *The Fifth Generation*, Signet.

Ferguson, C.H. & C.R. Morris 1994, *Computer Wars, the Fall of IBM and the Future of Global Technology*, Times Books.

Ferris, T. 1988, *Coming of Age in the Milky Way*, Doubleday.

Flanders, H. 1990, *Differential Forms with Applications to the Physical Sciences*, Dover.

Fukuyama, F. 1992, *The End of History and the Last Man*, Avon.

Gardner, J. 2011, *The Blitz*, Harper.

Gamota, G. & W. Frieman 1988, *Gaining Ground, Japan's Strides in Science & Technology*, Ballinger.

Gray, J. 2020, *Simply Riemann*, Simply Charly, Kindle.

Grove, A 1999, *Only the Paranoid Survive*, Currency.

Gullbert, A. 1997, *Mathematics, From the Birth of Numbers*, W. W Norton & Co.

Hardy, G.H. 1908, *A Course of Pure Mathematics*, Cambridge University Press.

Hardy, G.H. 1940, *A Mathematician's Apology*, Cambridge University Press.

Hargittai, I. 2006, *The Martians of Science, Five Physicists Who Changed the Twentieth Century*, Oxford University Press.

Hartwig, R.L. 2005, *Basic TV Technology*, Elsevier.

Hassig, L., ed. 1989, *Revolution in Science*, Time-Life Books.

Hellman, H. 2006, *Great Feuds in Mathematics*, Wiley.

Hoddeson, L. & V. Daitch, 2002, *True Genius, the Life and Science of John Bardeen*, Joseph Henry.

Horowitz, P. & W. Hill 1989, *The Art of Electronics*, Cambridge University Press.

Hsu, F.H. 2002, *Behind Deep Blue: Building the Computer that Defeated the Chess World*, Princeton University Press.

Hwang, K. 1984, *Computer Architecture and Parallel Processing*, McGraw-Hill.

Isaacson, W. 2005, *Kissinger, a Biography*, Simon & Schuster.

Isaacson, W. 2011, *Steve Jobs, a Biography*, Thorndike.

Jackson, T. 1997, *Inside Intel, Andy Grove and the Rise of World's Most Powerful Chip Company*, Dutton.

Johnstone, B. 1999, *We Were Burning, Japanese Entrepreneurs and the Forging of the Electronic Age*, Basic Books.

Kanigel, R. 1991, *The Man Who Knew Infinity, a Life of the Genius Ramanujan*, Scribner.

Kissinger, H. 1994, *Diplomacy*, Touchstone.

Komp, R.J. 2002, *Practical Photovoltaics, Electricity from Solar Cells*, AATEC.

Kuo, J.B. 1996, *CMOS Digital IC*, McGraw-Hill.

Lagrange, J.L. 1997, *Analytical Mechanics*, Springer-Science.

Lanczos, C. 1949 (1970), *The Variational Principles of Mechanics*, Dover.

Laplace, P.S. 1902, *A Philosophical Essay on Probabilities*, Wiley.

Lee, W.C.Y. 1989, *Mobile Cellular Telecommunications Systems*, McGraw-Hill.

Livio, M. 2005, *The Equation that Couldn't be Solved*, Simon & Schuster.

Margolis, A. 1985, *Computer User's Guide to Electronics*, Elsevier.

Marsden, J. & M.J. Hoffman, 1987, *Basic Complex Analysis* (2nd Edition), W.H. Freeman.

McNeill, D. & P. Freiberger 1993, *Fuzzy Logic*, Touchstone.

Mendelson, B. 1990, *Introduction to Topology* (3rd Edition), Dover.

Miller, A.I. 2019, *The Artist in the Machine, the World of AI-Powered Creativity*, MIT Press.

Millman, S., ed. 1984, *A History of Engineering and Science in the Bell System*, AT&T Bell Laboratories.

Minsky, M. & S. Papert 1969, *Perceptrons, an Introduction to Complex Geometry*, MIT Press.

Muller, N.J. 2001, *Bluetooth Demystified*, McGraw-Hill.

Muller, R.S. & T.I. Kamins 1986, *Device Electronics for Integrated Circuits* (2nd Edition), Wiley.

Murray, C.J. 1997, *The Supermen, the Story of Seymour Cray and the Technical Wizard Behind the Supercomputer*, Wiley.

Nahin, P.J. 2004, *When Least is Best*, Princeton University Press.

Nash, C. & S. Sen, 1983, *Topology and Geometry for Physicists*, Dover.

Nassar, S. 1998, *A Beautiful Mind*, Simon & Shuster.

Navidi, W. 2019, *Statistics for Scientists and Engineers*, McGraw-Hill Education.

Needham, J. 1954, *Science and Civilization in China* (Vol. I), Cambridge University Press.

Nielsen. M. 2018, *Neural Networks and Deep Learning*, Academia.edu.

Nietzsche, F. 1957, *The Use and Abuse of History*, Bobbs-Merrill.

Orton, J. 2004, *The Story of Semiconductors*, Oxford University Press.

Penrose, R. 1994, *Shadows of the Mind*, Oxford University Press.

Penrose, R. 2002, *The Emperor's New Mind: Concerning Computers, Minds & the Laws of Physics*, Oxford University Press.

Penrose, R. 2007, *Road to Reality: A Complete Guide to the Laws of the Universe*, Vintage.

Pierret, R.F. 1988, *Semiconductor Fundamentals* (Vol. I), Addison-Wesley.

Posamentier, S. & C. Spreitzer 2020, *The Lives and Works of 50 Famous Mathematicians*, Prometheus.

Proust, M. 2020, *Delphi Complete Works of Marcel Proust*, Kindle.

Quittner, J. & M. Slatalla, *Speeding the Net, the Inside Story of Netscape*, Atlantic Monthly Press.

Rabiner, L. & B.H. Juang, 1993, *Fundamentals of Speech Recognition*, Prentice-Hall.

Reid, C. 1996, *Hilbert*, Springer-Verlag.

Rhodes, R. 1986, *The Making of the Atomic Bomb*, Simon & Schuster.

Rhodes, R. 1995, *Dark Sun, the Making of the Hydrogen Bomb*, Simon & Schuster.

Riordan, M. & L. Hoddeson 1997, *Crystal Fire, the Invention of the Transistor and the Birth of the Information Age*, W. W Norton & Co.

Ross, S.M. 1970, *Applied Probability Models with Optimization Applications*, Dover.

Schewe, P.F. 2007, *The Grid*, Joseph Henry.

Shannon, C. & W. Weaver 1971, *A Mathematical Theory of Communication*, University of Illinois Press.

Shepard, S. 2000, *Telecommunications Convergence*, McGraw-Hill.

Slater, R. 1999, *Saving Big Blue*, McGraw-Hill.

Smith, D.E. 1959, *A Source Book in Mathematics*, Dover.

Stross, R.E. 1997, *The Microsoft Way*, Addison-Wesley.

Sze, S.M. ed. 1988, *VLSI Technology*, McGraw-Hill.

Szilard, L 1961, *The Voice of the Dolphins*, Simon & Schuster.

Tenenbaum, M. & H. Pollard 1985, *Ordinary Differential Equations*, Dover.

Thomas, A. 2013, *V1 Flying Bomb Aces*, Osprey Publishing.

Thomas, G.B. 1960, *Calculus and Analytic Geometry*, Addison-Wesley.

Tolstov, G.P. 1962, *Fourier Series*, Dover.

Transnational College of LEX 1995, *Who is Fourier, A Mathematical Adventure*, Language Research Foundation.

Van Zant, P. 1997, *Microchip Fabrication*, McGraw-Hill.

Virgil, *The Aeneid*, Book IV (fate of Queen Dido), Perseus Digital Library version of 1490 English translation.

Wallace, J. & J. Erickson 1993, *Hard Drive, Bill Gates and the Making of the Microsoft Empire*, Harper.

Weber, S. 2004, *The Success of Open Source*, Harvard University Press.

Weise, M. & D. Weynand 2007, *How Video Works, From Analog to High Definition*, Elsevier.

Whitehead, A.N. & B. Russell 1910, *Principia Mathematica*, Cambridge University Press.

Wiener, N. 1948, *Cybenetics: Or the Control and Communication in the Animal and Machine*, Martino Fine Books.

Wiener, N. 1950 (1988), *The Human Use of Human Beings*, Da Capo Press.

World Intellectual Property Organization 2019, *Technology Trends, Artificial Intelligence*, World Intellectual Property Organization.

Wozniak, S. 2006, *iWoz*, W.W. Norton & Co.

Index

Printed in the United States
by Baker & Taylor Publisher Services